RIGHT STUFF WRONG SEX

GENDER RELATIONS IN THE AMERICAN EXPERIENCE
Joan E. Cashin and Ronald G. Walters
Series Editors

Margaret S. Creighton and Lisa Norling, eds., *Iron Men, Wooden Women: Gender and Seafaring in the Atlantic World, 1700–1920*

Stephen M. Frank, *Life with Father: Parenthood and Masculinity in the Nineteenth-Century American North*

Lorri Glover, *All Our Relations: Blood Ties and Emotional Bonds among Early South Carolina Gentry*

Richard Godbeer, *Sexual Revolution in Early America*

Anya Jabour, *Marriage in the Early Republic: Elizabeth and William Wirt and the Companionate Ideal*

Angel Kwolek-Folland, *Engendering Business: Men and Women in the Corporate Office, 1870–1930*

Margaret A. Lowe, *Looking Good: College Women and Body Image, 1875–1930*

Catherine Gilbert Murdock, *Domesticating Drink: Women, Men, and Alcohol in America, 1870–1940*

Margaret A. Weitekamp, *Right Stuff, Wrong Sex: America's First Women in Space Program*

RIGHT STUFF WRONG SEX

America's First Women in Space Program

Margaret A. Weitekamp

THE JOHNS HOPKINS UNIVERSITY PRESS
Baltimore & London

Printed in the United States of America on acid-free paper
9 8 7 6 5 4 3 2 1

The Johns Hopkins University Press
2715 North Charles Street
Baltimore, Maryland 21218-4363
www.press.jhu.edu

Library of Congress Cataloging-in-Publication Data

Weitekamp, Margaret A., 1971–
Right stuff, wrong sex : America's first women in space program / Margaret A. Weitekamp.
p. cm. — (Gender relations in the American experience)
Includes bibliographical references and index.
ISBN 0-8018-7994-9 (hardcover : alk. paper)
1. Women astronauts—United States—Biography. 2. Women in astronautics—Political aspects. 3. Feminism. 4. United States. National Aeronautics and Space Administration—History. 5. Space race. 6. Sex discrimination against women. I. Title. II. Series.
TL789.85.A1W45 2004
629.45'0082'0973—dc22

2004008929

A catalog record for this book is available from the British Library.

To the memory of my father,

Raymond L. Weitekamp

CONTENTS

ACKNOWLEDGMENTS

When I began researching this history in 1996 as a dissertation for Cornell University's History Department, I was warned that it could not be done. Other people who had attempted to untangle the confusing threads of this story had given up in frustration. In addition, there existed no centralized archives of documentary sources to explain the program's shrouded beginning and prolonged end. Reconstructing the history of Lovelace's Woman in Space Program required personal trips to twelve states and a full year at the National Aeronautics and Space Administration (NASA) Headquarters History Office, courtesy of NASA and the American Historical Association. In the end, writing the book was an education in itself. People's generosity continually impressed me, and I am grateful for the chance to offer them some well-deserved recognition.

First of all, I must thank the remarkable women who told me their stories so that I could write this book. I could never have anticipated their hospitality. They put me up in their homes, took me to dinner, and showed me their lives. They carefully photocopied newspaper articles, photographs, and personal letters. I hope I have repaid their immense kindness by writing a narrative that does justice to the history they lived.

At Cornell University, I offer heartfelt thanks to Richard Polenberg. His guidance helped me to define the project and greatly improved both its content and its form. Thank you for unfailing support over many years. Margaret Rossiter offered critical readings and timely advice throughout the dissertation and since. Thank you also to Thomas Holloway and Thomas Borstelmann.

Winning the 1997 NASA/AHA Aerospace History Fellowship was a turning point in my career. Roger D. Launius, then the NASA historian, protected my workspace, heralded my project, and showed me opportunities that I never could have expected. Archivists Mark Kahn and Colin Fries happily found anything I needed and let me know about items I would never have discovered alone. Thanks also to Nadine Andreassen and Steve Garber. Together they make each trip back to the History Office both pleasant and productive. Thank you to Andrew Pedrick, Richard Faust, and the NASA Headquarters Library staff. Dorothy Cochrane, Valerie Neal, and Mike Neufield of the Smithsonian Institution's National Air and Space Museum each offered useful feedback. Thank you also to Andrew Butrica, Bettyann Holtzmann Kevles, Katherine Landdeck, and Jake W. Spidle Jr.

Generous financial assistance from several organizations supported my research and writing. I gratefully acknowledge the Mellon Foundation for the Humanities, the American Historical Association along with the National Aeronautics and Space Administration, the John F. Kennedy Presidential Library, the Lyndon Baines Johnson Presidential Library, Cornell University's Graduate School, and Cornell University's Women's Studies Department for financial support during the dissertation process. The 2002 Aviation/Space Writer's Award from the Smithsonian Institution's National Air and Space Museum supported the manuscript revision.

Special thanks to the staffs at the Kennedy Presidential Library, the Johnson Presidential Library, the Truman Presidential Library, the Eisenhower Presidential Library, and the Lovelace Sandia Health System. Thank you also to the staffs of the National Air and Space Museum, the University of New Mexico Medical School oral history collection, the Inhalation Toxicology Laboratory Library on Kirtland Air Force Base in New Mexico, and the Ninety-Nines International Organization of Women Pilots, Inc. Archivists at the University of Michigan and the Johnson Space Center also assisted me. The staffs of the Texas Woman's University WASP collection, the Margaret Chase Smith Library, and the Columbia University oral history collection sent valuable material.

At Hobart and William Smith Colleges, I thank Betty Bayer, Claudette Columbus, Jodi Dean, Lee Quinby, Susan Henking, Elena Ciletti, Lara Blanchard, Renee Monson, and the rest of the Not Piss Poor Women's Studies reading group for creating a challenging place to think critically about gender.

Anna Creadick, Kevin Dunn, and Bahar Davary deserve special thanks as valuable friends and colleagues.

Personal thanks to Gwendolyn Llewellyn, Emily Longnecker, the Myers-Hayes family, Kit Sturgeon, Kimberly Weinberg, Pamela Weinberg, Colleen and Ray Weitekamp, Jocelyn Waite, Michael Wolf, and the First Monday Book Club. Long overdue thanks to Deborah Layne and the University of Pittsburgh's Dean Alec Stewart. Thanks also to two wonderful teachers who pushed me as a writer: Terry Sturm and Maurine Greenwald.

Many thanks to Robert J. Brugger and Melody Herr at the Johns Hopkins University Press for their patience and guidance.

My parents, Ann and John Gannon, were unflagging in their daily support of this book. I cannot thank them enough. Finally, I must thank my husband, Kevin Michael Days. You know what it took to write this. Thank you for standing by me through it all.

RIGHT STUFF WRONG SEX

INTRODUCTION

On May 27, 1961, pilot Geraldine "Jerrie" Cobb, whom *Time* magazine had called "the first U.S. lady astronaut," posed alongside a full-scale model of the capsule designed to carry the first space travelers for the National Aeronautics and Space Administration (NASA). As she stood beside the seven-foot cone with "United States" emblazoned on its ridged metal hull, Cobb embodied the cultural and political debate in the early 1960s about whether women could be astronauts. Wearing high heels and white gloves, she smiled as the photographer captured her juxtaposed with the single-seat Mercury spacecraft. Two steps positioned by the hatch-turned-window allowed her and the other attendees at the First National Conference on Peaceful Uses of Space to peer inside. As a part of her campaign to become the first woman in space, Cobb posed with one foot resting on the top stair, as if ready to climb the final step into NASA's first spaceship.[1]

For almost a year, curiosity about Cobb's success on the Lovelace Foundation's astronaut tests (designed for NASA in 1958) had inspired demand for public appearances and continued media coverage of her actions. Attention to Cobb's achievements—and to Lovelace's ongoing women's testing program—would prompt NASA administrator James E. Webb to name Cobb as a special consultant to the U.S. space agency later that same evening. Yet on that May afternoon, even as Webb considered offering Cobb a consulting position, the capsule providing her photographic backdrop already contained an astronaut—a male mannequin lying on the ergonomic couch, dressed in a silver Project Mercury spacesuit and with his hand grasping the control stick. The pilot's seat was already full.

Cobb's publicity photo captured a telling moment when the U.S. space program, aerospace science, cold war politics, and gender relations intersected in a short-lived women's astronaut testing program. Throughout the spring and summer of 1961, Dr. William Randolph Lovelace II invited twenty-five women pilots to take his foundation's astronaut tests.[2] The privately funded program demonstrated that women would be well suited for space travel. By the time Cobb's campaign for a woman in space became public, however, the Mercury capsule already had a living occupant; NASA had its astronauts. By the end of the summer, Lovelace's initiative would be canceled, unable to continue without access to government-run aerospace facilities.

The United States had good reasons to consider a female astronaut. After all, the American space program had answered every other Soviet space advance. On October 4, 1957, when the Soviets launched *Sputnik*, the first earth-orbiting satellite, the United States rushed to match the achievement—at first failing publicly, when the navy's *Vanguard* satellite blew up on the launch pad, and finally succeeding with the launch of *Explorer 1* in 1958. Just a month before Cobb's photo opportunity, Yuri Gagarin became the first man to orbit the earth on April 12, 1961. NASA quickly launched astronauts Alan Shepard and Gus Grissom in separate suborbital flights to demonstrate that the United States could also put a human being into outer space—however briefly. John Glenn finally restored American parity with three orbits in February 1962. Yet when the Soviet space program launched Valentina Tereshkova into orbit as the first woman in space on June 17, 1963, the United States dismissed her feat as a publicity stunt.

Why did American decision makers consider the flight of a woman in space to be an achievement that did not require a response? Perhaps launching an American woman would signal that a direct competition for space supremacy existed. When NASA officials declared that the space agency pursued scientific achievements, not propaganda stunts, they took the United States out of the race for space firsts. At the time, the Soviet space program far outperformed the American effort in both lifting capacity and orbital duration. Tereshkova's flight highlighted the inequality; her total flight time surpassed all previous American launches combined. Refusing to participate in a race for firsts that had already been lost allowed the United States to save face. Even after President John F. Kennedy dedicated the American space program to a future moon landing, the space agency could refuse to engage in an explicit race to match each Soviet advance.

Space policy makers also feared that any harm to a female astronaut would outrage a public that remained protective of women. They feared that such a disaster could shatter the political support for the space program as a whole. The prospect of subjecting a woman to mortal danger betrayed the rigidly defined gender roles asserted in postwar America.[3] Given the military background of the leaders of the new civilian American space agency, many NASA officials simply could not conceive of women in the masculine role of astronaut.

At a very basic level, it never occurred to American decision makers to seriously consider a woman astronaut. In the late 1950s and early 1960s, NASA officials and other American space policy makers remained unconscious of the way their calculations implicitly incorporated postwar beliefs about men's and women's roles. Within the civilian space agency, the macho ethos of test piloting and military aviation survived intact. The tacit acceptance that military jet test pilots sometimes drank too much (and often drove too fast) complemented the expectation that women wore gloves and high heels—and did not fly spaceships.

Rigidly defined gender roles also played out in the realm of foreign policy. In the early 1960s, American policy makers evaluated spaceflight options by assessing whether they provided appropriately strong responses to Soviet advances. Flying a member of the "weaker sex" into space evoked the old aviation stereotype—usually wrong but nonetheless widespread—that if a woman could fly a craft, it must be easy to do. Indeed, Western space experts cited Tereshkova's mission as evidence that anyone could occupy the automated *Vostok* capsules. In contrast, the arguments went, America's Mercury spacecraft required skilled astronaut pilots with engineering backgrounds and experience in test flying military jets. The very qualifications required for NASA astronauts proved the complexity of U.S. space achievements. Demonstrating that a woman could perform those tasks would diminish their prestige.

Besides, American women already had a role in the cold war mobilization: a domestic one. In the often-cited 1959 "kitchen debate," when Vice President Richard Nixon showed Soviet premier Nikita Khrushchev how the appliances in the model kitchen at the American National Exhibition in Moscow eased life for U.S. housewives, Khrushchev countered that Soviet women contributed to the USSR as productive workers. American ideology depicted Soviet women as masculine, defeminized by their work roles and lack of domesticity. In contrast, women advanced the United States' cause by maintaining the American way of life at home. The projection of women's domestic roles into international competition peaked when Tereshkova's flight earned approval while Lovelace's women encountered significant resistance.[4]

As a result, a balanced historical analysis of Lovelace's female astronaut testing project requires understanding how gender operated, not as a single point of analysis or focus of inquiry, but as an inherent, inextricable element of many levels of historical analysis. In the United States in the early 1960s, gender pervaded aerospace medicine, popular culture, domestic politics, and foreign policy.[5]

Despite the many levels at which gender operated, an awareness of how it shaped American culture had only just begun to emerge in the early 1960s. The era when Lovelace's Woman in Space Program began and ended, the late 1950s and early 1960s, has been called a time of "prefeminist agitation" on the eve of the modern women's movement. In 1961, Women Strike for Peace organized fifty thousand housewives to walk out of their homes to protest aboveground nuclear testing, and President John F. Kennedy appointed former first lady Eleanor Roosevelt to head the Commission on the Status of Women. Two years later, when space policy makers faced Tereshkova's launch, Betty Friedan published *The Feminine Mystique* and Congress passed the Equal Pay Act.[6]

By 1964, the disrupting potential of women's unchecked power would also emerge as a theme in popular television shows. Situation comedies such as *Bewitched* and *I Dream of Jeannie* featured women whose extraordinary powers upset men's lives but whose actions could always be neutralized in the end. In *I Dream of Jeannie*, NASA itself was the masculine hierarchy that fell apart when a woman intruded. Years before Jeannie disoriented Major Nelson, however, Jerrie Cobb and the women of Lovelace's Woman in Space Program presented a real-life challenge to the social and political order used to justify and organize American space efforts.[7]

The public debate over Lovelace's Woman in Space Program linked the cultural turmoil surrounding women's changing roles to policy making at the national and international levels. Although the events of the early 1960s excited public curiosity about women astronauts, when the opportunity arose to answer Tereshkova's flight, postwar cultural baggage outweighed budding prefeminist agitation. No sustained movement for women's equality existed to provide a political impetus for putting a woman in space.

As a program encompassing American space exploration, aerospace science, cold war politics, and gender relations, Lovelace's Woman in Space Program offers much more than just the intriguing tale of a project that began twice and then lingered for two years after it had been canceled. Lovelace's women's testing project illuminates a moment when an unexpected program briefly took flight. It was launched by researchers excited by the boundless possibilities of space exploration and kept aloft longer than anyone thought possible by those ready to believe in women's capabilities. That the program ended should not be surprising given the forces arrayed against it. Rather, that it existed at all reveals how aerospace medicine, cultural politics, and gender relations intersected when the early stirrings of the women's movement coincided with the development of the United States' *manned* space program.

“GOING TO TOWN FOR THE MEN OF SCIENCE”

1

Randy Lovelace and Jackie Cochran

Just after noon on December 17, 1940, three of the United States' top aviation doctors and the nation's most famous woman pilot gathered in the White House to attend President Franklin Delano Roosevelt's presentation of the Collier Trophy for the previous year's outstanding aviation achievement. Roosevelt wanted the ceremony to be small, preferably no more than ten people, and the negotiations about who should attend had taken several weeks. The three named award recipients—Dr. Walter M. Boothby, Dr. William Randolph Lovelace II, and Army doctor Capt. Harry Armstrong—would be present, of course. The question had been whether there would be room for their benefactors and mentors: Dr. Charles Mayo of the Mayo Clinic in Rochester, Minnesota, and Miss Jacqueline Cochran, a famous woman pilot and a member of the Collier Trophy selection committee. White House permission for the two guests came through only one week before the ceremony.[1]

As the small group filed into the office on that day, the dramatic Collier Trophy was already there. When the publisher Robert J. Collier, an early aviator, commissioned the prize in 1911, the sculptor responded with a massive piece portraying the drama of flight. The polished wooden pedestal alone measured almost a foot high and a foot and a half square. On each of the four sides, inscribed plaques listed the award's purpose and the previous honorees. Above this corniced base rose a two-foot-tall bronze sculpture of three figures and a bird swirling around a globe. The largest one, a stylized male figure, rose from the orb with clouds at his feet. The bronze tribute to the triumph of human flight rested heavily on the president's desk, weighing fully 525 pounds.[2]

As the photographers began to take pictures, the newest recipients of the Collier Trophy gathered behind FDR's chair to be photographed with the president and the trophy. Roosevelt rested his hand on the award's base, smiled for the cameras, and chatted with the physicians. The picture FDR later autographed for Dr. Lovelace showed the three doctors and the president posing behind the award.[3]

Another photograph taken that day, however, shows more of the room and tells more of the story. In that picture Boothby, Lovelace, and Armstrong still cluster behind the president. Off to the left, however, Cochran peers out from the mammoth trophy's shadow, crowded into the background. She is only partially visible—barely in the picture. Cochran stands in the background, however, not because others crowded her out of the picture but because she pushed to be in it. Indeed, without her none of these men would have been at the awards ceremony at all. In a calculated campaign orchestrated from within the Collier Trophy committee itself, Cochran arranged for the three aviation doctors to receive the 1939 award. As the guiding force behind this White House gathering, she made sure that the photographers recorded her presence.[4]

Cochran campaigned for her preferred candidates to receive the award for the outstanding aviation achievement of 1939 in order to demonstrate that she could exercise influence on a national level. With the European war looming and German air power sweeping across the English Channel during the late summer of 1940, Cochran wanted the United States to pay attention to its own air corps. As she explained it, all other aeronautic advances depended on aviation medicine. By bringing the doctors' work to the Collier Trophy committee's attention, she brought national recognition to medical advances in high-altitude flying and helped to prepare her country for the coming war. In doing so, she solidified her reputation as an influential figure and forged a lifelong friendship with a young aviation researcher.

The story of how Jackie Cochran helped Randy Lovelace win the 1939 Collier Trophy is the story of a budding personal and professional association that lasted a lifetime. Furthermore, it is the story of aviation's booming development and expansion. Only years after barnstorming introduced the nation to flying, men of science expanded the possibilities of human flight and daring women fliers tested those limits. Most significantly, however, the story forms the first chapter in the history of American women's quest for space.

Understanding how a program for potential female astronaut candidates came to exist in 1960 requires knowing the backgrounds and personalities of the two major actors who created it and sponsored it. By pushing Lovelace into the spotlight—and herself into the national political picture—Cochran cemented a friendship that set the stage for the pair to collaborate on Lovelace's Woman in Space Program in the early 1960s. Twenty years after the Collier Trophy awards ceremony, Lovelace's vision and Cochran's support created an opportunity for women to test their mettle as potential astronauts.

Randy Lovelace and Aviation Medicine

Randy Lovelace's invitation to the White House to receive the nation's highest aviation award came only a year after he completed his formal medical training as a surgeon. Indeed, as a doctor, Lovelace found his lifelong passion early. In the 1930s, as a part of a cohort of research physicians working in the armed forces and at research centers such as the Mayo Clinic, Lovelace helped pioneer the specialty of aviation medicine—the new science of measuring and regulating the environmental stresses pilots encountered—even as he finished his own medical education. His impressive professional development grew out of his close family upbringing. As a young man, the mentoring of the uncle he was named for guided his educational choices and steered him toward cutting-edge aeromedical research.

Born December 30, 1907, in Springfield, Missouri, William Randolph "Randy" Lovelace II moved to New Mexico as a toddler when his family relocated to care for Randy's uncle and namesake, William Randolph Lovelace. The elder Lovelace had contracted tuberculosis in early 1906, less than a year after completing medical school. Seeking a cure in New Mexico's high, dry climate, the new Dr. Lovelace joined the Santa Fe railroad as a company surgeon. When he relapsed, however, his parents and sisters came there to care for him. Randy's parents joined the family in 1908. Randy Lovelace thus spent his early years surrounded by extended family amid the rugged New Mexico landscape. In 1913 the entire family moved to Albuquerque. When Randy's parents divorced in 1918, the eleven-year-old boy went to live with his uncle and grandparents.[5]

For the rest of his life, "Uncle Doc" (as the elder William Randolph Lovelace came to be known, even to his colleagues) became a second father to Randy. Whenever Randy made major decisions, he consulted both his father and Uncle Doc. After Randy earned a private pilot's license at Washington University in St. Louis, he wrote home for advice when the navy ROTC offered him advanced flight training in 1928. Although his father consented, Uncle Doc did not approve, and Randy declined the opportunity. Instead of becoming a military pilot, he finished college and enrolled in Washington University's medical school. Growing up around his uncle's medical practice directly influenced Randy's decision to pursue medicine. Loving and demanding, Uncle Doc offered advice—sometimes heeded, sometimes not—throughout a medical school career that included two years at Washington University, a year at Cornell, and a final transfer to Harvard. In 1934 Randy graduated from Harvard University's medical school, becoming the second Dr. Lovelace in the family.[6]

Uncle Doc also introduced Randy to the institution that set the young doctor's career on the fast track: the Mayo Clinic. Since their first meeting with Uncle Doc in 1915, Dr. William Mayo and Dr. Charles Mayo, the brothers who founded the Rochester, Minnesota, institution, profoundly shaped Lovelace's Albuquerque medical practice. Uncle Doc patterned his own medical partnership, the Lovelace Clinic, after the Mayo Clinic's group practice model. When the Lovelace Clinic recruited physicians, many of the new doctors came directly from Mayo. In 1934, after much urging, Uncle Doc persuaded Randy and his new wife, the former Mary Moulton, to visit the Mayo Clinic. They loved it. By the next Christmas Randy began a surgery fellowship, and in 1936 the couple moved to Minnesota, where they spent the next six years.[7]

At Mayo, Lovelace achieved his expertise in the fields that defined his career: surgery and aviation medicine. While studying surgery, he also became the student of Dr. Walter M. Boothby, a physiologist engaged in pioneering medical research into flight's effects on the human body. To pursue this interest, Lovelace took a brief leave in 1937 to train as a flight surgeon at the School of Aviation Medicine at Randolph Field in Texas. (Dating back to World War I, military practice dubbed all doctors who screened and monitored pilots' health "flight surgeons" whether or not they specialized in surgery.) Back at Mayo, Lovelace became a significant collaborator in Boothby's investigations.[8]

Throughout the 1930s, flight surgeons searched for solutions to the problems pilots encountered as advancing commercial and experimental flying subjected them to new stresses—ones that low-altitude fliers had never encountered. Reliable new aircraft spawned commercial businesses. During the 1920s and into the 1930s, commercial airlines flew the mail, shipped freight, and transported passengers. At the same time, races and attempts to set speed and altitude records tested the capabilities of both planes and pilots. Commercial pilots, air racers, and record setters alike experienced problems created by the limitations on their own physical performance.

As advancing aircraft designs allowed airplanes to fly ever higher, oxygen deprivation caused medical symptoms that compromised pilots' ability to fly. Reactions slowed. Judgment lapsed. Even worse, pilots sometimes behaved irrationally. As the oxygen in the atmosphere thinned with altitude, a pilot could make fundamental errors—heading east instead of west or up rather than down—without knowing the difference. Rather than descending to a safer altitude, the pilot might continue to climb, worsening the condition. As a result, such "pilot errors" were compounded because pilots did not realize that they were no longer functioning normally. Physical symptoms usually arose long after muddled judgment did, and at higher altitudes. Eventually pilots became lightheaded, felt their skin crawl, or experienced pain in their joints. They could even black out. Long before blackouts occurred, however, they faced real danger. As an editorial from the 1938 *Annals of Internal Medicine* warned, "When traveling at 200 miles an hour even a slowing of the mental processes, with consequent increase in the so-called 'reaction time' is extremely dangerous, especially if the weather is bad, visibility poor and 'instrument flying' necessary." In fast new airplanes, even momentary lapses of control could prove deadly.[9]

Commercial aviation faced a different, but related, set of medical problems. As air transport ventures developed, their innovators increasingly looked to higher altitudes for faster, more economical travel. Oxygen deprivation symptoms exacted a new cost when passengers entered the aircraft. Beyond the obvious risks of pilot incapacitation during a flight, passenger discomfort could be materially dangerous and financially disastrous. Rapid altitude changes unsettled passengers. Commercial airlines also needed to alleviate the noise, smell, vibration, and temperature problems that made early air travel uncomfortable. Commercial aviation could not continue to advance without figuring out how to keep both pilots and passengers comfortable and alert at high altitudes.[10]

Throughout 1938, Lovelace and Boothby searched actively for solutions to these aeromedical problems. To find a remedy, they joined a military-sponsored project to develop a usable oxygen mask for high-altitude flight. The Army Air Corps hired the two Mayo physicians as dollar-a-year consultants, granting them access to the facilities at the Aero Medical Laboratory at Wright Field.[11]

While at Wright Field, the two doctors tackled the altitude puzzle by conducting experiments using a specially built low-pressure chamber. The device allowed them to simulate the lower atmospheric pressure that pilots encountered as they flew higher. Northwest Airlines volunteered its pilots to take turns in the pressure chamber. When using similar equipment at the Mayo Clinic, the doctors enlisted local high-school boys to undergo treadmill fatigue tests in the low-pressure chamber. Boothby and Lovelace also experimented on themselves. The results allowed them to isolate the medical problems of high altitude.[12]

Building on earlier research that linked oxygen deficiency to pilot error, the Mayo doctors identified two problems concerning blood gases and altitude. First, in the rarefied air at high altitudes, a pilot could not easily breathe enough oxygen to remain lucid. Furthermore, the absence of normal atmospheric pressure hindered the blood's oxygen absorption, exacerbating symptoms. Oxygen insufficiency resulted. Second, without sufficient atmospheric pressure, nitrogen gas came out of solution in the pilot's blood, causing symptoms of the "bends" or caisson sickness. At high altitude a pilot's blood "boiled."[13]

The reliable portable oxygen mask that Boothby and Lovelace developed at Wright Field in 1938 offered a solution. Since oxygen deprivation produced the symptoms, a steady supply of the essential gas would make high-altitude flying safer. Before their innovation, pilots had sucked supplemental oxygen from a large tank through a tube attached to pipestem held between their teeth: a cumbersome and inefficient system. The Mayo doctors offered a mask that combined reliability with comfort. At Wright Field, Dr. Arthur Bulbulian, a prosthetist who specialized in constructing artificial noses, helped the design team contour the mask to fit snugly over the user's face. The physicians called the final result the BLB mask, after the initials of its three creators: Boothby, Lovelace, and Bulbulian. The oxygen delivery system came in two models: a lighter version intended for use by airline passengers and a heavier version designed for military and competition flying. Both models consisted of a small portable oxygen tank connected to a face mask. The lighter model covered only the nose, framing the user's mouth in an open O that allowed for comfortable eating and talking while aloft.[14]

The national significance of their oxygen research and the BLB mask became apparent very quickly. Even as early as September 1940, the U.S. Army had already ordered six thousand of the portable oxygen systems. In years to come, the BLB mask became standard issue for high-altitude flying during World War II. The aeromedical research that Boothby and Lovelace conducted at Wright Field with Harry Armstrong, who headed up the Wright Field medical labs, led to their commendation by the Collier Trophy award committee in 1940. But though their work eventually saved many lives, brilliant research alone did not win them the prestigious Collier Trophy. The pioneering doctors came to the attention of the selection committee because of the efforts of a famous woman pilot: Jacqueline Cochran.[15]

Jackie Cochran and Women's Aviation

Cochran was a self-made woman in the most literal sense. As a poor but ambitious young girl, she envisioned the life she wanted—one with riches and success—then worked for the rest of her life to live that vision, as a record-setting pilot, a successful business owner, an international celebrity, and a friend to generals and presidents. A natural politician, she negotiated circles of powerful men using her status as "the only" and "the first" woman to gain access to influence normally denied to women at that time. She could be a formidable opponent or an invaluable ally. By her own account, she started from nothing, gaining everything she had or was through her own persistence. Forceful, determined, outspoken, glamorous, and vivacious, Cochran embodied a larger-than-life persona of her own creation.

What is known about Jackie Cochran's early life came entirely from her own recollections and stories passed along by her friends. In her 1954 autobiography, *The Stars at Noon*, written when she was approximately forty-eight, she described her childhood living with a foster family in a series of northern Florida sawmill towns. Because she had little formal education, she struggled with reading and writing throughout her life. From an early age, however, the smart and articulate girl became determined to forge a new path for herself.[16]

In a rags-to-riches story, Cochran carefully crafted a public persona, along with a new name to match her new life. Living on her own in Pensacola, Florida, during her late teens, she decided she needed a new last name to complete her new independent identity. She chose "Cochran" from a phone book. As Jacqueline Cochran, she made a final break with her previous life, declaring her separation from a poverty-stricken childhood where she had never felt wanted. Cochran lived the rest of her life on her own terms.[17]

Although information about her biological family was available to her, she never felt she needed it. When she was about to marry, she worried that her soon-to-be husband, Floyd Odlum, might want to know about her heritage, so she obtained "sealed letters from the two people then living who might know the facts" of her early life and real parentage. Her husband returned the envelopes unopened. They remained in Cochran's locked box, still sealed, for the next four decades. When Odlum died in 1976, Cochran's friend Chuck Yeager burned the sealed notes in her presence.[18]

As a young woman, the new Miss Cochran used her skills as a beautician to finance her personal and social advancement. Having learned hairdressing as a live-in maid, Cochran began assisting in a beauty shop, giving permanent waves. As her clientele grew, so did her ambitions. When a traveling salesman came into the shop looking for an expert operator to run a permanent wave machine in Montgomery, Alabama, Cochran recalled, "I thought this would be a good way to start my travels. I walked into the room and said I was that expert." In Montgomery, one of her clients encouraged her to continue to push herself. After a few unsuccessful attempts at selling dress patterns door-to-door throughout the South and teaching beauty school in Philadelphia, Cochran finally established herself as a beautician at a posh New York City salon, Antoine's.[19]

There in New York City in 1932, Cochran encountered the two great loves of her life: Floyd Odlum and flying. At the time, Odlum ranked as one of the United States' richest men. He ran the Atlas Corporation, a financial giant that made fortunes buying and selling utilities companies. Although he was married when he met Cochran at a dinner party, their friendship quickly blossomed. They married in 1936. The same night she met Odlum, she began thinking about aviation. In her autobiography Cochran recalled, "I think I got the idea about flying that first night during the [dinner] table conversation with Floyd." Rather than driving up and down the East Coast following her wealthy clients, he suggested she get a private pilot's license and become a flying beauty operator and cosmetics agent. Cochran took three weeks' vacation at Long Island's Roosevelt Field to learn to fly. According to *The Stars at Noon*, at a time when it usually took several months to earn a pilot's license, Cochran soloed on her third day, completing her training before her vacation ended.[20]

Getting a private license did not satisfy Cochran. As a pilot—as in everything else she did—Cochran pushed herself, supremely confident in her own abilities and convinced that she could do whatever she wanted to do. Within weeks of learning to fly, she set off for Montreal to attend an air meet without knowing how to read a compass, let alone how to navigate with instruments. She later used her connections with several naval officers she knew from Pensacola to learn to fly "the navy way" at a California naval base. Within two years of her first lesson, she entered the world of air racing.[21]

During the 1930s, highly publicized cross-country and transcontinental air races captured the public's attention by featuring the stars of the aviation world. In 1929 Clifford Henderson began the National Air Races, a collection of aviation competitions held each year. Starting in 1931, the Bendix Corporation also sponsored a prestigious transcontinental air race offering a substantial purse. Although not all races welcomed female entrants during the early 1930s, by the end of the decade women competed side by side with men in most events. Women also set separate speed and altitude records. Since the Fédération Aeronautique Internationale maintained a distinct category for women's aviation records, women received publicity for setting official marks even when they did not surpass the absolute record for the category, a boon for their publicists and sponsors. Although Cochran entered this competitive world slightly later than many of her contemporaries, she quickly made her presence felt.[22]

During the 1930s, Cochran gained notoriety as a successful woman air race pilot who preferred to compete with men. In fact, despite a widespread circuit of all-women air races that thrived in the 1930s, Cochran's only entrance into an all-women race (a pylon race, in which the competitors flew around high obstacles) came in her preparation for the 1934 MacRobertson London-to-Australia air race.[23] Throughout her life as an aviator, Cochran chose racing and interacting with men over participating in women's events, preferring to be the first or only woman to accomplish an aviation feat. As a result, throughout her career, Cochran's efforts to promote women remained secondary to her desire to be aviation's exceptional woman.

Cochran thrived at a time when women pilots did receive national attention yet still faced biases in the world of aviation. Air race organizers hesitated to include women pilots and overreacted to women's accidents. In 1933, when Florence Klingensmith died during the Frank Phillips Trophy Race in Chicago, organizers used her death to justify barring women from the 1934 Bendix race even though the cause of Klingensmith's accident—the airplane's wing covering tore apart at high speed—would have killed any pilot, male or female. Cochran helped women earn their way back into national air meets in 1935, only three years after she learned to fly. She lobbied until she could enter the Bendix. However, as one of the first two female entrants (along with her friend Amelia Earhart), she had to obtain waivers from all the male pilots as a condition of her entry. She withdrew because of engine problems, but her admittance reopened the door for Louise Thaden and Blanche Noyes to win the Bendix the next year.[24]

In the late 1930s, Cochran came into her own as an air racer and aviation record setter. About 1938, her relationship with Alexander de Seversky, a Russian-born aircraft developer, gave her the means to move to the front of the pack. Seversky needed to demonstrate his new designs so that he could sell airplanes to the United States military. When skeptics credited his long-distance speed to his own piloting skill and not to the aircraft's design, he turned to Cochran. Flying Seversky's aircraft, she promptly set a new women's speed record and "an absolute record for a flight from New York to Miami." In 1938, she finally conquered the Bendix. Flying a Seversky P-35 pursuit, she achieved the second female Bendix victory, handily defeating a field of nine men. By 1940 she had won the Harmon Trophy as the outstanding woman flier for three years running. By the decade's end, Cochran's attainments put her in an exclusive category of aviation's elite, both male and female.[25]

She fit the category comfortably. Cochran rose to prominence in a cohort of accomplished female fliers whose characteristics she shared. Women pilots of the 1930s typically were affluent enough to afford flying competitively, even in the midst of the Depression. Many famous female pilots used personal wealth, rich husbands or promoters, and industry patrons to finance their aviation. For Cochran, a combination of Floyd Odlum's money and her own tenacity opened doors. When Cochran wanted to learn to fly, Odlum paid for lessons. When she wanted to race, she persuaded airplane manufacturers to give her airplanes to fly.

Aircraft manufacturers used women pilots to demonstrate the safety and ease of their new prototypes. When Seversky asked Cochran to fly his new airplanes, her flights not only tested the plane's capabilities under race conditions but also showed that the craft could perform exceptionally, no matter what. If a woman could handle the new design, the reasoning went, then surely the army's military aviators would have no problems. In truth these women, and Cochran in particular, were highly skilled pilots. For the 1934 London-to-Australia race, Cochran obtained a Granville Brothers Gee Bee racer, a fast but unstable aircraft that killed many pilots. Cochran remained one of the few pilots to fly that airplane successfully for any length of time. Nonetheless, as Cochran recalled, the perception persisted in aviation that "if a woman can fly it, anyone can."[26]

As much as Cochran knew that her flying accomplishments could help Seversky because she was female, she remained supremely conscious of how her image as a woman pilot advanced her own interests. Whether or not they employed the aid of a publicist, female pilots during the 1930s knew that publicity was the key to additional sponsorship and therefore more challenging flying opportunities. Women fliers had to strike a delicate balance between the heavy flight jackets and trousers they wore while accomplishing public feats and the feminine appearance sponsors wanted to see. Cochran took this balancing act one step further, using public expectations for how a lady should be treated as a means to advance her ambitions.

Cochran's Public Face

Cochran built a reputation as a woman who excelled at what she did and generally got what she wanted. A driven woman with a clear vision of her goals, she made her way in the world by combining purposeful femininity with good old-fashioned political smarts. As an aviator, businesswoman, and celebrity, Cochran maintained a carefully constructed image as a fashion plate, well coifed and impeccably made up, who could fly with the best pilots in the world. In contrast, Cochran's contemporary, the much-studied Amelia Earhart, challenged gender stereotypes through her actions and appearance. Whereas Earhart encouraged women to work together for change, Cochran understood the power that a woman gained from being the only female in a room. To make the most of that advantage, Cochran used the unwritten but well understood rules of etiquette—the conventions for how a lady should be treated in polite company—to negotiate her place in a man's world.

Cochran counted herself a close friend of Earhart, the famous pilot and early feminist who became infamous after her 1937 disappearance while trying to set a round-the-world record. In the months before her final ill-fated flight, Earhart stayed at Cochran and Odlum's California ranch to rest and prepare. Indeed, after the disappearance, Cochran tried to locate her lost friend using psychic intuition. Cochran dabbled in spiritualism, considering herself to have a well-developed sixth sense. She and Earhart even experimented with trying to locate downed aircraft together before Earhart left for her round-the-world trip. Not being able to sense her lost friend's location greatly disappointed Cochran.[27] Beyond their friendship and their common status as accomplished aviators, however, Earhart and Cochran differed greatly in their attitudes about women's status and their personal philosophies about appearance, relationships, and advancement.

Through her flying accomplishments, outspoken advocacy, and physical appearance, Earhart acted in the spirit of the women's movement, even as the

1920s and 1930s saw a marked decline in active feminism. She publicly supported feminist causes, aligning herself with women's professional organizations such as Zonta, International, and endorsing the Equal Rights Amendment in a telegram released at the National Women's Party's 1936 convention. In her personal attire and presentation, Earhart so successfully blended feminine touches with the masculine hallmarks of a pilot—a flight jacket and trousers—that the look became her trademark. Her easily recognized image combined a short hairstyle, a leather flight jacket, and tailored trousers with a ladylike silk blouse or strand of pearls. Earhart's stylish assertion of individualistic talent set an example for women in the immediate postsuffrage era.[28]

Like Earhart, Cochran kept her own name after she married, acting as a public figure independent of her husband. She entered into marriage with less reluctance than Earhart did, however. Fearing the loss of freedom that being someone's wife would entail, Earhart outlined a famous prenuptial agreement, asking George Putnam to grant her a divorce in a year if their marriage did not bring them both happiness. In contrast, Cochran maintained her individual identity as a public person while embracing her marital partnership with Odlum (and all the personal and professional benefits that accrued from it). The way she used her name illustrated how she saw the relationship. Although she wanted to be addressed as Miss Jacqueline Cochran in all professional correspondence, she expected formal social invitations to be addressed to Mrs. Floyd Odlum.[29]

Cochran delighted in her marriage to Odlum. By all accounts, theirs was a lifelong romance and a unique partnership. When chronic arthritis frequently left Odlum racked with almost crippling pain, Cochran tended him with great affection. As a friend of the couple recalled, "It was true devotion and lovely to see." For his part, Odlum did whatever he could to make Cochran's wishes come true, even arranging at one point for her to fulfill a childhood dream of riding an elephant in a circus. Although busy schedules often kept them apart, they remained close, acting in tandem to manage their social lives and her professional concerns. Since Cochran's near illiteracy hampered her correspondence, Odlum worked out a system to help her. After Cochran dictated what she wanted to say, he helped her put it into formal language, writing out the final letters in longhand for her on yellow legal pads for secretaries to type. Throughout their married life, Odlum supported her business and aviation ambitions, both emotionally and financially. For Earhart and Cochran, their husbands—Putnam the promoter and Odlum the financier—materially advanced their flying careers.[30]

In contrast to descriptions of Earhart as a postsuffrage feminist, Cochran would be better characterized as the sole female member of the old-boys network. She excelled at using personal connections, professional friendships, and business contacts aggressively to advance her causes. Her personal appearance became a key component of this strategy.

Unlike Earhart, whose public image paired a leather flight jacket with a silk scarf, Cochran favored glamorously feminine dress. She loved fashionable clothes, completing her look with makeup of her own creation. Throughout her life she had long blond hair, usually worn pulled back from her face and curled. According to her secretary, "People used to call her the Golden Girl because she was always sort of golden. Her skin was tanned, her hair was blonde, and she just kind of shone wherever she went." Cochran worked hard to maintain her look no matter what the circumstances. Even after her grueling record-setting flights, she recalled, "I always tried when getting out of my plane, no matter how hard the trip, to look as nearly as possible as if I had just stepped out a bandbox."[31]

Cochran used her attention to personal appearance as a way of asserting her importance. After winning the 1938 Bendix, she made the race committee wait while she fixed her face. As she recalled in her autobiography, "When I landed at Cleveland after winning that race, one of the judges came out to the end of the runway by car to pick me up and take me to the platform in front of the great crowd of people present. He had to wait while I sat in my plane and tidied up my appearance with comb and make-up. He looked quite disgusted when he saw what was going on." No matter how upset the race official became, however, he knew that good manners forbade him from yelling at a woman in public or openly badgering her to rush. Cochran knew those boundaries too and used them to her advantage. As a woman, she drew power from the unwritten rules of chivalry.[32]

Cochran regularly appealed to the social protections of being female. On another such occasion, after she crash-landed in Bucharest, she "refused to get out of the plane until I had removed my flying suit and used my cosmetic kit." The image of Cochran applying makeup while still wearing her flight gear became so characteristic that the photograph chosen to accompany her 1940 *Current Biography* entry shows her seated in the cockpit of her airplane, wearing her leather flying helmet and goggles, applying lipstick with the aid of a small mirror. For Cochran, the deliberately paced grooming after the 1938 Bendix race and the Bucharest crash was part of a pattern of behavior. Keeping people waiting allowed her to assert power. More specifically, keeping *men* waiting while she fussed with her face in public—something that only a woman could do—allowed her to command attention by forcing powerful men to defer to her. Cochran used such tactics not only to draw attention to herself as the exceptional woman in a man's world, but also to exercise influence in a way that could not be easily countered by any man who considered himself polite.[33]

For Cochran, applying makeup in public also did double duty as a business promotion. Since she was the head of Jacqueline Cochran Cosmetics, makeup and appearance were literally her business. From the very beginning of her business, when Odlum helped her set up the enterprise with his financial backing, Cochran involved herself in all levels of the venture, mixing many of her own formulas and personally selecting colors. As a public person, her dual roles reinforced each other. Cochran's aviation records advertised her cosmetics business, and her business dealings promoted her record-setting flights. For both ventures, Cochran regularly graced newspapers and magazines in photographs and interviews. Whether in her aviation career or her cosmetics business, Cochran managed her image deftly.

Throughout her career, in her dealings with military generals, airplane manufacturers, elected politicians, and other male public figures, Cochran used her conspicuously ladylike appearance to manipulate men's gallantry. She was not above flirting a little with a powerful man—teasing him that she had caught him staring at her legs, for instance—to disarm him by flattering his ego. Since her method of operating depended on being the only woman in a situation, she seldom went out of her way to encourage women's success generally. From the powerful men she operated with, Cochran also learned how to maximize personal and professional connections in order to advance her career and theirs.

When Cochran met Dr. Lovelace for the first time at the 1937 National Air Races in Cleveland, she recognized that he could help solve her aviation oxygen problems. Cochran's interest in supplemental oxygen had grown out of perilous firsthand experience. In 1937 she attained an altitude record of 33,000 feet in an unpressurized, fabric-covered biplane while sucking oxygen through a pipestem attached to a large oxygen tank. Despite using the supplemental gas, the inadequate delivery system left her fuzzy headed. As she recalled, "I became so disoriented through lack of oxygen that it took over an hour to get my bearings and make a landing." When she finally reached the ground, she discovered that her nose was bleeding. The low atmospheric pressure had ruptured a blood vessel in her sinuses. Inadequate oxygen needlessly increased the danger of her flight.[34]

When Cochran learned about the research being conducted by the Mayo Clinic doctors, including the development of a reliable oxygen mask, she immediately recognized how significantly the innovation could change the experience of high-altitude flying. As a pilot herself, she could testify to the need for improved oxygen delivery systems. Furthermore, she recognized how professionally advantageous it would be for her to bring the innovation to the attention of aviation's elite. In the 1940s, Cochran used her status as a famous 1930s "aviatrix" and an international celebrity to launch herself to a position of national influence.[35]

Cochran's Campaign

In June 1940 the National Aeronautic Association (NAA) assembled a list of aviation's most notable figures to serve on the committee charged with selecting the Collier Trophy winner for 1939. From the award's inception, the Collier Trophy recognized "the greatest achievement in aviation in the United States, the value of which has been demonstrated in actual use during the preceding year." In 1940 the select group invited to determine what accomplishment most deserved the trophy's recognition included Maj. Gen. H. H. "Hap" Arnold, chief of the Army Air Corps, as well as Eastern Airlines president Capt. Edward V. "Eddie" Rickenbacker. Robert E. Gross, president and chairman of the board of Lockheed Aircraft Corporation, and Dr. George Lewis, director of the National Advisory Committee for Aeronautics, also accepted the invitation. The NAA committee also solicited no less than Grover Loening, the pioneering aircraft designer and previous Collier Trophy winner, who had learned to fly from Orville Wright himself and then directed Pan American Airways. Cochran's inclusion on this exclusive invitation list reflected her standing as one of the biggest names in aviation.[36]

The Collier Trophy's prestige also attracted high-powered applicants for the award. The nomination process itself became competitive. Both United Air Lines and Hamilton Standard Propellers took it upon themselves to tout their recent achievements in formal presentations to the selection panel. In addition to those suggestions presented to the committee, each award committee member could submit recommendations. Among the twenty-two possible award recipients nominated at the first meeting, Cochran offered five suggestions. Aviation medicine topped her list.[37]

Like those of the other committee members, Cochran's assessment of the previous year in aviation reflected the growing war being waged on the ground in Europe and in the air over Great Britain. As the committee surveyed 1939's aviation achievements, the Luftwaffe's ferocious dive-bombing during the Nazis' sweep across Europe remained fresh in everyone's mind. Even more so, only three weeks before their first meeting on July 10, 1940, the Battle of Britain had begun as a series of skirmishes over the English Channel. While Cochran wrote an open letter to the entire Collier Trophy selection committee in early August 1940, German air forces escalated their attacks on the Royal Air Force. Because the front lines of the European battle had risen to the skies above Great Britain, Cochran's thoughts turned to American air power.

Any assessment of aviation's role in the U.S. arsenal, however, would not prove commendable. In correspondence with the committee members, Cochran outlined the difficulties of presenting the Collier award for the state of American national defense: "We haven't achieved or accomplished a strong air defense, nor did France, Belgium, Holland, Norway or Poland." Indeed, she pointed out, "If a trophy were to be given to the one who proved in 1939 in actual demonstration that a strong air force or air defense was necessary, it would have to be given to Hitler."[38] With the prospect of the United States' entry into the war looming larger by the day, Cochran wanted American military aviation to receive more attention.

Nominating an air power advocate risked politicizing the award, however. The foremost spokesman for a strong air defense remained Gen. Billy Mitchell, the Air Services officer whose outspokenness on the issue provoked his famous 1925 court-martial. In 1940 the debate about whether aviation required a separate military branch remained contentious. Cochran realized that suggesting Mitchell as a Collier Trophy candidate, even after his death, threatened to turn the committee vote into a referendum on a separate air corps. Even worse, awarding the trophy to Mitchell could make it appear that the NAA endorsed military reorganization. The Collier Trophy was not intended as a political statement.[39]

Dismissing the national defense choice as untenable, Cochran argued strongly for two alternatives involving aviation safety: give the award either to commend high-altitude oxygen deprivation researchers or to recognize the commercial airlines for their outstanding 1939 safety record. For both selections, she suggested adding a special citation for "Doctors Lovelace and Boothby of the Mayo Clinic and Captain Armstrong of Wright Field." The doctors' achievements not only had been "demonstrated in actual use" in the previous year as required by the trophy mandate but also, in Cochran's opinion, held the key to future wartime flying. As she reminded her fellow committee members, "Air warfare as now conducted, depends greatly for its effectiveness on proper application of oxygen." She saw aviation medicine making advances that would help the United States achieve parity with advanced German technologies. Awarding the trophy according to her suggestions would reward a worthy achievement while also advancing American preparations for war.[40]

Furthermore, as Cochran presented it to the committee, much of 1939's safe commercial flying could be attributed to better medical understanding of oxygen deficiency. She argued, "I believe that the reduction of accidents was due primarily to reduction of so called 'pilot error.' . . . Lack of a clear head it has been now proven is really in most cases brain fatigue due to oxygen deficiency." Cochran knew the significance of the doctors' achievements from her own high-performance flying. Now she set out to convince the Collier Trophy committee of the importance of these "unknown men."[41]

When Cochran decided she wanted something, she marshaled all of her considerable resources to achieve it. For the Collier Trophy selection process, she wrote letters, solicited testimonials, and lobbied friends. She also corresponded actively with her physician nominees, advising them about how to manage their submissions to the selection committee. Cochran threw herself into the project enthusiastically. As she informed other aviation colleagues more than once, "I am 'going to town' for the men of science and have hopes of success."[42]

Cochran had only a few weeks before the late September final vote to convince her fellow committee members that three virtually unknown aeronautical researchers deserved the nation's highest aviation award. Throughout September 1940, Cochran increased her campaign for Boothby, Lovelace, and Armstrong. She solicited letters from both the president and the chief test pilot of Consolidated Aircraft Company attesting to the doctors' work. Also at her request, Frank Fuller Jr., a 1938 and 1939 Bendix air race competitor, wrote a letter to the committee describing how well his oxygen equipment had functioned during the cross-country races and what a difference it had made for him.[43]

The Mayo Clinic doctors enthusiastically prepared research briefs for the committee and consulted with Cochran about how to improve their chances most effectively. She had been in contact with them even before they had officially been nominated. Two days before she sent her open nomination letter to the Collier Trophy committee, Cochran wired Dr. Lovelace at the Mayo Clinic, asking him to send her sample masks and supporting data about the equipment. By the time the awards committee requested additional materials, she had already organized a well-thought-out promotional campaign to support the doctors' nomination.[44]

The doctors also consulted Cochran's husband, Floyd Odlum, about the most effective way to approach the award committee. When Boothby and Lovelace received Odlum's letter informing them that their nomination had officially reached the next round of consideration, they each had the same idea. In separate letters, they both offered the same suggestion. They could transport a small low-pressure chamber to Washington, DC, and demonstrate their findings to the committee in person. As Boothby effused, "This makes quite a striking demonstration, especially as he is able to go up in ten minutes, after being decompressed, and can come down again in less than ten minutes." Boothby even suggested that they might allow the committee members to experience the pressure chamber themselves, if it was convenient. He wondered if Odlum thought a live demonstration of the oxygen equipment would help their case.[45]

Working as a team with Odlum, Cochran replied to the doctors' proposal and advised them about further submissions. She warned them that bringing the cumbersome apparatus to Washington, DC, would cause more trouble than benefit in the long run. Instead, she urged them to work on compiling their brief for formal submission. She advised the research physicians to obtain testimonials from the major airlines using the BLB masks, reasoning that since the airlines were also still in contention for the Collier Trophy (for their safety record), they should be happy to cooperate in their own interests. At her suggestion, Boothby and Lovelace also had reprints of their published works bound into a volume to accompany the brief.[46]

As she prepared to present the doctors' case to the committee, Cochran personally supervised the distribution of their supporting materials. She organized all the letters and testimonials they had sent to her and prepared a packet for each of the committee members, topped with a four-page cover letter of her own endorsing the researchers. Cochran stressed the physicians' excellence and their unselfish service to the cause of aviation. As she had when she nominated them, Cochran also pointed out the necessity of this research for the European war effort, reminding the committee members that "the findings have indeed contributed no end to the effective night work that the British Air Forces have been carrying out over Germany." In fact, she added, Capt. Harry Armstrong, the head of Wright Field and a nominee along with Boothby and Lovelace, had recently been called to Canada to help the British conduct similar studies on high-altitude oxygen use.[47]

In her thoroughgoing campaign, Cochran omitted nothing. In Cochran's hands, Armstrong's new publication became a weighty handbill for her campaign. In the growing field of aviation medicine, Capt. Harry Armstrong had literally written the book on the subject. His newly published *Principles and Practice of Aviation Medicine* outlined the extensive research findings compiled at Wright Field. In case any members of the Collier Trophy committee had missed Armstrong's authoritative new text, Cochran arranged for Brentano's Book Stores to deliver a copy to each one a week before the final voting meeting.[48]

Cochran also used her personal contacts to influence her colleagues and assess the committee's direction before the final vote. She arranged for friends of hers to call on those committee members she did not know well. One of them, Cochran's friend Harry Bruno, reported back to her that he had talked with Frank Tichenor about the Collier Trophy nominations. By mid-September, when she had not gotten a response from the materials she sent to Maj. Gen. Hap Arnold, she asked a general she knew to speak to Arnold. As befitted such an organized election campaign, Cochran started counting the probable votes well before the decisive meeting. She wanted to make sure that favorable delegates would attend so they could support her candidates. Unsure whether he would attend, she wrote to Eastern Airlines president Eddie Rickenbacker offering to transmit his vote to the committee if he could not be present at the voting meeting.[49] She left no possible contingency unexamined.

Cochran lobbied the committee effectively. By the time she attended the decisive award selection meeting on September 29, 1940, her doctors remained as one of only three nominees still on the agenda. One of the other possibilities offered for consideration also came from Cochran's original five suggestions: recognizing the commercial airlines for their safety record. In the end, the committee awarded the 1939 Collier Trophy to both of Cochran's remaining nominees. The commendation inscribed on the award's brass plaque recognized both the airlines' safety record and the aviation doctors.[50]

The magnitude of the award sank in slowly for Lovelace and Boothby. Upon hearing the good news, the two doctors wired Cochran thanking her politely and expressing the Mayo Foundation's "deep appreciation" for her help. Their initial formality belied their excitement, however. Thrilled and grateful, they thanked Cochran profusely for her actions on their behalf. Five hours after the first telegram, as it neared 11:30 p.m., they cabled again, jubilantly, "The four of us are drinking your health and that of your husband in a series of bottles of Mumms." More than just toasting her proposal of their nomination or their good fortune in winning, they acknowledged that Cochran's well-organized campaign and personal clout had earned them the committee's favorable votes. As Lovelace wrote to her four days later, "I do not believe any man on the committee would have stuffed his guns like you did." They recognized that their acclaim resulted directly from Cochran's determined advocacy.[51]

Overnight, the aviation doctors' research into oxygen use became widespread news. The press lauded the Collier Trophy committee's choice with great fanfare. As Frederick Graham opined in the *New York Times*, "It would be difficult, if not impossible to imagine more appropriate recipients for the 1940 Collier Trophy Award than those named last week." And, as was customary for the Collier Trophy winner, in addition to the general news coverage of the award ceremony, *Collier's* magazine ran a detailed story explaining the doctors' achievements to the general public. Readers learned about the insidious effects of oxygen deficiency and how the doctors were helping to make the skies safer. Accompanying the story, color photographs showed Boothby and Lovelace in their pressure chamber and airline passengers dining comfortably while wearing BLB masks. The article emphasized how thankful its readers would be for the many ways that the doctors' research benefited the flying public.[52]

At the time, however, the greatest gratitude came from Dr. Lovelace and his colleagues for Cochran's extraordinary advocacy. Winning the Collier Trophy only six years after receiving his MD (and before reaching his thirty-third birthday) established Lovelace as one of the premier experts in aviation medicine. In truth, the Collier Trophy selection process also benefited Cochran tremendously. Her participation allowed her to strike up an acquaintance with Maj. Gen. Hap Arnold, the chief of the Army Air Corps, a man who would prove to be a strong ally in her next great undertaking: mobilizing women pilots during wartime. Perhaps more significant, the 1939 Collier Trophy selection process began a lifelong personal and professional relationship between Cochran and Lovelace. Twenty years after they had their photograph taken together in the White House, that friendship led Cochran to offer financial support for another Lovelace research project: the first systematic testing of American women as potential astronaut candidates.

But before they could work together on that experiment, aviation medicine had to evolve into aerospace medicine and women's aviation had to produce a broader candidate pool than the "society-girl pilots" of the 1930s. World War II accelerated both of these changes. During the war, Germany introduced practical rocketry to the world, beginning a race that extended the boundaries of flight into the stratosphere and challenged flight surgeons to assess human adaptability in harsh spacelike conditions. The exigencies of wartime also expanded female pilots' opportunities for the duration and redefined women's aviation in the postwar period. In rising to these challenges, the next episodes of Lovelace's and Cochran's lives laid the immediate groundwork for Lovelace's Woman in Space Program.

2

"THIS BUCK ROGERS NONSENSE"

Aviation and Aerospace Medicine

On June 24, 1943, a B-17 Flying Fortress droned high above Ephrata, Washington. Sitting inside the bomb bay, Lt. Col. Randy Lovelace adjusted his face mask and double-checked the oxygen tank strapped to his left leg. He felt nervous, and he had good reason. He had never jumped out of an airplane before. Nonetheless, he was preparing to make the highest parachute jump ever attempted. If all went well, when he touched the ground he would not only hold the world record for the parachute jump but would also have proved that his portable oxygen system was reliable for high-altitude bailouts.[1]

Lovelace's personal testing of the invention's reliability demonstrated a common practice in aviation medicine. As he recalled years later, "It [the mask] was perfected in the laboratory. However, you always miss out on something in the laboratory, and you have to put the equipment to use to find out if it is successful." Lovelace and his cohorts regularly subjected themselves to ascents and descents in pressure chambers to test their equipment. Lovelace's life-threatening jump simply took this self-experimentation one step further. His actions exemplified the practices of a cohort of aviation research physicians who explored the limits of human physiology from the 1940s into the 1950s.[2]

As they solved the medical problems facing pilots at high altitudes, aviation researchers also began to think about the medical problems of flight in outer space. They faced two obstacles. They needed to figure out how to protect a human being from the harsh environment of outer space. At the same time, they needed to negotiate a different hostile environment: the one found at the highest ranks of the military and in Washington, DC. Government and military officials derided interest in human spaceflight as akin to dime novel science fiction—perhaps an interesting sideline for a private researcher, but certainly an irresponsible use of the taxpayers' money. In addition to conducting their actual research, then, budding aerospace experimenters also had to protect their experiments from superiors opposed to spending money on space research.

In the late 1940s and 1950s, Lovelace participated actively in organizing conferences and conducting experiments that changed aviation medicine into aerospace medicine. As they expanded the field, researchers reconceptualized the upper atmosphere, developed a new understanding of high altitude's effects on the human body, and found methods of simulating those effects in a laboratory. The last step required new technologies. As aerospace medicine developed during the 1950s, medical researchers increasingly relied on simulation apparatus, diagnostic machines, and protective accessories to probe human reactions to stratospheric altitudes. Lovelace played a central role in this evolution, both through his individual research and through the namesake research foundation he created in Albuquerque.[3]

By the end of the decade, a lot had changed. When NASA's first astronaut candidates traveled to the Lovelace Foundation in 1959 to be vetted for space travel, aerospace medicine was a thriving field, the government was sponsoring a crash program to put a human being into orbit, and the name "Lovelace" evoked the cutting edge in aerospace physical testing. From World War II until the late 1950s, however, Randy Lovelace worked to solve the medical problems of human spaceflight in the face of government officials who actively discouraged such goals.

The Jump

After winning the 1939 Collier Trophy, Dr. Lovelace continued to investigate pilots' oxygen needs as he was promoted to colonel in the U.S. Army Air Corps and chief of the Oxygen Branch at the Aero Medical Laboratory at Wright Field in Dayton, Ohio. In 1941 the U.S. military issued the BLB (Boothby-Lovelace-Bulbulian) mask to all pilots for high-altitude flying. By 1943, however, wartime situations posed new challenges for the standard-issue apparatus.

When gunfire damaged the oxygen systems in combat aircraft, pilots sometimes blacked out from oxygen deprivation before they could climb out of the cockpit. A supplemental oxygen tank might buy an aviator the chance to parachute out of a doomed airplane. But how should the pilot proceed when forced to bail out at high altitude? Should he open his parachute, using the supplemental oxygen as he floated down into breathable air at a lower altitude? Or should he free-fall to that safer altitude, minimizing his time in rarefied air before deploying his chute? The army needed to know, and there was only one way to find out. Lovelace decided to test his invention personally in a world-record parachute jump.[4]

As Lovelace checked his equipment for the last time, the B-17 climbed over 40,000 feet. To protect himself from the bitter cold outside, Lovelace wore layers of the army's high-altitude flying gear. Over his pants and shirt, he donned an insulating "chicken-feather suit" with a thick fur collar. He also covered his hands in layers, wearing a pair of thin rayon gloves under his heavy aviator gloves. His gabardine outer jumpsuit bulged beneath the straps of the parachute harness. His parachute would open automatically.[5]

Finally, the aircraft reached 40,200 feet. The War Department's report recorded the rest of the experiment in eloquent detail:

> Colonel Lovelace climbed down onto a seat in the bomb bay to face the rear, leaned forward, and turned on the bail-out oxygen supply. Miles away he saw the Columbia River, a thin, shining ribbon twisting through fields of green. Thirty thousand feet below, a liaison plane moved along its course, ready to follow the jumper to earth. As his plane sped forward at 200 miles per hour, Colonel Lovelace stepped off into air 50° below zero. The tail of the plane was directly over his head when the 'chute started to open. The blast of onrushing air and the sudden jerk of the opening parachute tore the thick outer glove from both his hands and also snapped off the thin inner glove on his left hand, leaving it bare in freezing temperature. The inner glove remained on his right hand, which was uninjured by the cold.[6]

The "sudden jerk of the opening parachute" also knocked Lovelace out cold. Unconscious, he plummeted 10,000 feet through the air, swinging helplessly below the parachute. Finally, at 8,000 feet, the oxygen flowing through the BLB mask revived him. Weak from the ordeal, Lovelace landed roughly, dragging behind the parachute before finally coming to rest.[7]

The experiment succeeded, but just barely. Lovelace and his colleagues had drastically miscalculated the physics of his unprecedented jump. As Lovelace recalled years later, "We assumed that the opening shock would be less up there than at low altitudes, so we thought that we'd open the parachute immediately. That's what I did. It turned out that was a mistake." Instead of encountering minimal air resistance, the force of opening Lovelace's parachute at high altitude not only knocked him unconscious but began to tear his parachute's tethers loose from their moorings. An estimated thirty or forty g's of force bent the harness's metal fixtures.[8]

The experience taught Lovelace and his colleagues a pointed lesson about how little they understood high-altitude conditions. When the Army Air Forces (AAF) researchers realized what Lovelace had encountered, they significantly redesigned high-altitude protective clothing and parachutes. Furthermore, the AAF ordered its pilots to postpone opening their parachutes in emergency high-altitude bailouts until they reached safer altitudes. In recognition of Lovelace's contributions, Maj. Gen. H. H. "Hap" Arnold awarded him the Distinguished Flying Cross.[9]

In the AAF, Lovelace's wartime work put him in the company of some of the best researchers in the country. As aviation medicine developed from the 1930s through the 1950s, the military provided an important training ground. The services offered the resources needed to simulate flight conditions using expensive apparatus. Several important research centers boasted advanced diagnostic and simulation technologies as well as innovative leaders. In 1935 Wright Field became the discipline's center when Harry Armstrong became the chief of the Physiological Research Laboratory. After the war, the air force organized the Air Research and Development Command (ARDC) as a site for aviation research and development, including aviation medicine. Military influence reached beyond formal research centers, however. Because the war effort required so many flight surgeons, many scientists who pursued aeromedical research during the postwar era had backgrounds that included military aviation medicine. The combination of medical schooling, aeronautics training, and military practice shaped how they approached their work.[10]

As physicians accustomed to working in an engineering environment, aeromedical researchers approached the human body as an engineering problem. In any flying machine, the "human factor" remained the one component that could not be reengineered, redesigned, refined, or improved. Aviation's technological development would always be constrained by the pilot's physical abilities and limitations. In the quest to protect pilots and improve aviation, aeromedical physicians tested and retested their subjects. They worked to establish the human body's thresholds for motion, vibration, altitude, oxygen, acceleration, and deceleration.

In their search for human subjects to experiment on, researchers often volunteered themselves. After all, the test designers knew what reactions to anticipate as well as what was unexpected. They also knew the potential risks best. Capt. Harry Armstrong, who shared the Collier Trophy with Lovelace in 1939, used himself as a subject in the Wright human centrifuge and later experimented on himself again regarding explosive decompression in the Lockheed XC-35. Likewise, Lovelace offered himself as an experimental guinea pig more than once. In addition to his parachute jump, Lovelace participated as both researcher and test subject in explosive decompression experiments in pressure chambers. Armstrong and Lovelace typified a cohort of military men who advanced the field of aviation medicine through experiments that sometimes risked their own lives.[11]

Although Lovelace later joked that his world-record parachute jump seemed to overshadow his other credentials ("I must have done something else worth mentioning," he quipped), his wartime aviation experimentation served as a significant precursor to the research that explored the boundaries of outer space. When the war ended, like many other aviation doctors who supported America's fighting pilots, he returned to civilian practice. In the years immediately after World War II, Lovelace founded a research center in Albuquerque that provided a new site for advanced aeromedical experimentation outside the military. Sadly, however, the impetus came from personal loss.[12]

On leaving active service, Lovelace returned to his promising surgical practice with Dr. Charles Mayo at the famous Mayo Clinic in Rochester, Minnesota. He, his wife Mary, and their three children settled happily in Rochester, where his career flourished. Then tragedy struck. In 1946 the Lovelaces' two sons developed polio. One after the other, the boys died. The Lovelaces were devastated. In their grief, they moved their household back to Albuquerque, New Mexico, where Randy joined his uncle's medical partnership, the Lovelace Clinic. Although they had considered returning to Albuquerque, they probably would not have moved so soon if not for the loss of their sons.[13]

In order for Randy to pursue his interests within the new partnership, Uncle Doc agreed to help him reorganize the clinic. The elder Dr. Lovelace remained committed to personal attention in primary care. Randy Lovelace envisioned bringing the Mayo Clinic's research and education to the Southwest. Together they founded the Lovelace Foundation for Medical Education and

Research in 1947 as an equal partnership where each man could pursue his interests. While Uncle Doc maintained a clinic focused on quality treatment, Randy created a research center based explicitly on the Rochester, Minnesota, model: a Mayo Clinic of the West. The two Lovelace doctors transferred the complete assets of the Lovelace Clinic to support their new endeavors, a financial commitment that required expert advice.[14]

Randy Lovelace turned to a man he had met during the Collier Trophy campaign. Floyd Odlum, the prominent financier, Jackie Cochran's husband, and Lovelace's old friend, agreed to become chairman of the board of directors. Through the early years of the Lovelace Medical Center, Odlum's directorship guided the Foundation to financial security. At the same time, Randy Lovelace developed the medical center's research and teaching, creating fellowships for medical study. The Foundation began attracting medical residents and research contracts.[15]

Throughout the late 1940s and into the 1950s, the Lovelace Foundation developed a national reputation. One early addition to the research staff, Dr. Sam White, won a Douglas Aircraft Corporation contract to study how carbon dioxide built up in airplane cockpits. The Foundation began performing physical examinations for the pilots of several major airlines. By 1951 it also handled significant aeromedical research projects for the U.S. Air Force. Lovelace's continuing relationship with air force researchers (including his position on the service's Science Advisory Board) kept him in touch with scientists whose research extended to the edges of the atmosphere and beyond. Pursuing those interests solidified the Foundation's reputation as an organization in the vanguard of what soon became aero*space* research.[16]

The "Upper Atmosphere"

In 1951, with the help of the Air Force School of Aviation Medicine, the Lovelace Foundation organized the first of two pathbreaking conferences that transformed aviation medicine into aerospace medicine. Although the logistical requirements challenged the young institution, the successful meeting fundamentally reshaped the field. By redefining how researchers understood the very structure of the atmosphere, the findings presented opened the door for investigations of human spaceflight. Despite having such significant research on the schedule, however, the organizers proceeded cautiously. Because the environment for space-related research remained hostile among military and government officials, they organized the symposium so that it would not seem too "spacey."

Serious research into how the human body could withstand the extreme conditions of outer space began only a few years after the end of World War II. As early as 1948, Capt. Harry Armstrong convened a panel called "Medical Problems of Space Travel." As the head of the Air Force School of Aviation Medicine, Armstrong established a Department of Space Medicine in 1949, appointing Hubertus Strughold to direct it. But this early research, conducted openly and discussed broadly, soon met with opposition.[17]

Fearful of an economic downturn as the wartime economy made the transition back to peacetime production, government officials hesitated to spend money on research into otherworldly pursuits such as space travel. Although resistance to space research existed well before 1952, under President Dwight D. Eisenhower an especially hostile environment took hold. Eisenhower's secretary of defense, "Engine" Charlie Wilson, took budget reductions seriously. In response to his repeated urgings, the Eisenhower administration favored limited defense spending. Wilson dismissed investigations into the nature of outer space as frivolous. In fact, as Lovelace physician A. H. Schwichtenberg recalled, when Secretary Wilson found out that several military doctors "were working with what he called 'this Buck Rogers nonsense,' . . . he called the three services' Surgeon Generals in to the Office of the Secretary of Defense and [really read them the riot act]." Interested researchers quickly learned to hide their projects from skeptical superiors.[18]

Aviation physicians became adept at disguising space-related research within other projects so they could work undisturbed. Instead of giving up their experiments, Schwichtenberg remembered that doctors at the Lovelace Foundation "decided to put all the money that was involved in those activities about three or four layers deeper in the budget." Since Secretary Wilson missed the announcements of the Foundation's successful biological tests—several of which involved sending small laboratory animals up in rockets—they got away with it. Likewise, Gen. Donald Flickinger of the air force rearranged the ARDC budget to disguise space-related experiments. Rather than give up on projects facing high-level resistance, aviation researchers downplayed the work's space-related implications—or outright space goals.[19]

The Lovelace Foundation doctors helped organize the 1951 conference in order to allow like-minded scientists to share what they had learned about the feasibility of human space travel. Well aware of the hostile environment for such research, the organizers deliberately left the word "space" out of the conference title. Instead, Lovelace and his colleagues at the Lovelace Foundation and the Air Force School of Aviation Medicine advertised the meeting as a conference about the "upper atmosphere." Even after four hundred attendees witnessed landmark discussions with clear ramifications for space exploration, the conference's published proceedings emphasized aviation applications rather than extraterrestrial possibilities. But the spin put on the results mattered less than the newest revelation: what could be learned within the atmosphere had direct implications for space. It all began with a new definition.[20]

At the 1951 conference Hubertus Strughold, the "father of space medicine," extended the limits of aviation medicine by introducing a new definition of "aeropause," the understood delineation between the earth's atmosphere and outer space. Rather than defining outer space as beginning at a boundary of six hundred miles altitude, Strughold asserted that "the transition from atmosphere to space technology takes place, not along a sharp topographically defined line, but rather in steps or stages within a very broad range, depending on the functions of the atmosphere." Simply put, spacelike conditions existed within the earth's atmosphere. Because conditions changed as the atmosphere thinned, the environmental stresses that a human traveler encountered also increased in stages. As a result, Strughold announced, medical researchers should think about the atmosphere as a continuum leading to space rather than as a sharp divide from outer space.[21]

Strughold offered an elegant diagram to illustrate his point. From the earth's surface to 120 miles high, Strughold charted atmospheric conditions of gradually increasing "partial space equivalence." Between 120 miles and 600 miles (the "material limit" of space), he labeled a region of "total space equivalence," which he called the "aeropause." As Strughold explained, his findings meant that "we might speak of biological *space-equivalent altitudes* within the physical atmosphere." Strughold's insight greatly simplified the work of researchers curious about human survival in space. Understanding how a biological specimen would react to space conditions no longer required sending it 600 miles high; space experiments could be done at specific altitudes *within* the earth's atmosphere. As a result, "upper atmosphere" experiments suddenly equaled "space-equivalent" research. After 1951, the physicians who conducted such experiments became known as aerospace researchers.[22]

The very term *aerospace* recognized the new understanding of the continuum between aviation and spaceflight. Although an altitude of fifty miles became the designated dividing line, no real boundary existed between the earth's atmosphere and outer space. (In fact, atmospheric gases can still be detected over one hundred miles above the earth's surface.) Between 1951 and 1958, American experimenters slowly figured out the risks of human existence in the aeropause.[23]

Determining how a person fared under high-altitude conditions required either sending someone into "space-equivalent" altitudes or simulating the experience in a laboratory. Not surprisingly, several researchers volunteered themselves as test subjects. For instance, John Paul Stapp, who had gained national attention (and a 1954 *Collier's* magazine cover) for his rapid acceleration and deceleration experiments using rocket-propelled sleds, joined forces with Joseph W. Kittinger Jr., another aerospace researcher with a bent for self-experimentation, on the Manhigh balloon project. Together they used specially designed balloons to ascend to high altitudes in order to test human reactions to spacelike conditions. In 1957, Kittinger extended the experiments by spending seven hours sealed in a pure oxygen environment gondola attached to a balloon. (The gondola reached 96,000 feet, spending two hours above 92,000 feet.) Kittinger also built on Stapp's acceleration runs using free-fall experiments. In 1960 he broke his own record with a seventeen-mile free fall from 103,000 feet (before landing safely by parachute). Experimenters also used simulation technologies to recreate potential space conditions while remaining safely on the ground.[24]

The need for rocket sleds, pressure chambers, and other expensive equipment concentrated aerospace research at the few centers with the resources to provide such apparatus. At Wright Field, the equipment available even included a full-sized human centrifuge. Established laboratories such as those at Wright Field and the Mayo Clinic maintained their superb reputations by attracting research fellows and project funding. As aerospace research developed in the 1950s, the Lovelace Foundation in Albuquerque, New Mexico, emerged as another site where medical experts worked with pressure chambers and simulation technologies to unlock the secrets that the upper atmosphere held about outer space.

After organizing the 1951 conference, the Lovelace Foundation continued to build its reputation as a prominent center for both aviation and aerospace medicine. The addition to the staff of Dr. Walter Boothby, Lovelace's mentor and fellow Collier Trophy recipient, further enhanced the Foundation's clout. The Lovelace Foundation became well known as a center for both research and practice. Throughout the mid-1950s, the Foundation's reputation, its new facilities, and Randy Lovelace's charisma attracted medical doctors and aerospace researchers from across the country.

The chance to work with Lovelace on his pathbreaking research outweighed the spartan conditions those new doctors found in Albuquerque. Dr. Donald Kilgore, a family practitioner who signed on with the Lovelace Foundation after completing a stint in the navy, recalled his wife's reaction when he first introduced her to their new hometown. "We came down U.S. 66 from Tijeras Canyon out here and in the far distance we could see this dusty little town. There was no green anywhere. [And she thought,] 'What have you done? Why are we here?'" It was certainly not the pay that brought the nation's top physicians to relocate to what Kilgore called "the brownest, dustiest place in the country."[25] Uncle Doc kept salaries low: adequate for Albuquerque's cost of living but certainly not nationally competitive.

Nonetheless, the opportunity to work on leading-edge aerospace medicine with the top researchers in the field drew many talented young doctors to Albuquerque. Lovelace's energy and vision were the magnet that attracted such talent despite the isolated setting. As Kilgore described it, "Randy's particular ability was to gather people up in the excitement of his ideas and to sweep them along. He got all kinds of people, including me, to come out here because of that rare ability."[26]

The logistical success of the 1951 conference notwithstanding, Lovelace's expertise lay in inspiration, not administration. As Ulrich Luft, the head of the Lovelace Foundation's Physiology Department, remembered him, "He was a great leader but I don't believe he was that much involved in administration and minute details. He was more a man that saw the grand design." Lovelace relied on the staff he assembled to turn his ambitious ideas into realistic programs. Dr. Robert Secrest remembered with a laugh, "Randy was a great one for saying, 'Don't tell me how difficult it is. Just tell me when you get it done.'" As Lovelace's first-rate staff successfully got jobs done, the Foundation began to handle even more work.[27]

Research funding flowed in steadily. In 1951 the Lovelace Foundation won an Atomic Energy Commission (AEC) contract to study the effects of nuclear explosion shock waves on biological specimens. As the Foundation demonstrated its ability to handle complex government research, Lovelace and his doctors landed a major government project: the top-secret air force contract to support the pilots for the U-2 spy planes. In 1955 the Lovelace clinic provided initial examinations. After they passed through security clearance, the Strategic Air Command's top pilots flew to Albuquerque for complete physicals. The testing regimen determined which of them physically qualified for the classified program. In addition, the Lovelace Foundation took responsibility for monitoring the pilots' health throughout their assignment to the U-2 project.[28]

The thoroughness of the Lovelace tests earned the respect of government supervisors and the exasperation of the medical subjects. The Lovelace physicians needed to prescreen the pilot candidates, protect them at high altitudes, and keep them clear-headed through long hours of grueling stratospheric flying. Testing and monitoring the U-2 pilots called on everything the Lovelace Foundation physicians had learned about the upper atmosphere. To guarantee the pilots' performance in this extreme environment, Foundation researchers carried out exacting physical examinations. For the pilots who participated, the exams' invasiveness left an uncomfortable impression. But the government and military officials assigned to monitor the Foundation's work were impressed with the Lovelace researchers' ability to work in partnership with U.S. government agencies, even under the added pressures of secrecy.

Although the U-2 contract with the Lovelace Foundation indirectly provided government funding for space-equivalent research, the atmosphere for space-related projects remained hostile. Even as late as February 1957 (after the United States declared that it would enter an artificial satellite as its contribution to the international scientific exhibition called the International Geophysical Year), Gen. Bernard Schriever received a rebuke from the Pentagon for his public interest in outer space. As head of the Western Development Division (renamed the Ballistic Missile Division later that year), Schriever gave a speech emphasizing the importance of space supremacy for national security. The next day he received a wire from the Pentagon requesting that he not use the word *space* in his speeches.[29]

Dr. Simon Ramo, another aerospace pioneer, reported a similar incident. Two days after the Soviet Union launched the first artificial satellite, *Sputnik 1*, on October 4, 1957, he changed the name of his private firm's systems engineering and technical direction division to the "space technology" unit. He received a call from the Pentagon the next day. The military official expressed fear that the name change would be perceived as a sign that Ramo's "feet were not on the ground."[30] Despite the pressure, Ramo did not alter the new title.

He was confidant that a sea change had occurred, putting his division in the forefront of a new enterprise. Within the next year, the official entry of the United States into the space race proved him right.

NASA's de Facto Medical Department

During the first few years after *Sputnik*'s launch announced the beginning of the space age, the direction and form of the U.S. effort was widely contested. While space boosters clamored for the United States to become a leader in human spaceflight—fulfilling the dream of having people living and working in space—other scientists argued that methodical experimentation should take precedence over piloted missions. Politicians and public figures either urged a crash program to match Soviet advances or cautioned that such action would waste money and resources. As decision makers negotiated the demands of competing constituencies to create a program and a policy, the early U.S. space effort took several years to take form.

For instance, space exploration did not immediately find a centralized home. In 1958 the military services maintained separate space programs. In addition to their ballistic missile projects and plans for launching artificial satellites, several of the services also announced human spaceflight programs. The air force named its project (a proposal to use ballistic missiles for a piloted suborbital flight) the Man in Space (MIS) program. After *Sputnik*, it became Man in Space Soonest (MISS) to emphasize its urgency. Werner von Braun's group, the Army Ballistic Missile Agency (ABMA), proposed a similar plan called Project Adam. The ABMA aimed to achieve suborbital space travel using the army's separate store of rockets. The navy suggested an even more ambitious step: orbiting a human being using all new technology, a 1958 project dubbed Manned Earth Reconnaissance (MER). All three plans developed simultaneously, and each hoped to succeed first.[31]

In fact, in the wake of *Sputnik*, the various U.S. military programs seemed to be competing with each other more than with the Soviets. On January 5, 1958, the *Milwaukee Journal* published a political cartoon depicting high-ranking members from each military service perched side by side on a cross-beam high atop a totem pole. The pole resembled a rocket and was labeled "Missile Program." The caption, "Each One is High Man on the Totem Pole," lampooned the military branches' misguided rivalry.[32]

Six months later, President Eisenhower moved to end such interservice bickering by creating one centralized space agency. Eisenhower's initial attempt to downplay *Sputnik*'s significance disguised his interest in allowing the Soviet Union to establish the strategic principle of peaceful overflight. That same interest in keeping space peaceful led Eisenhower to insist on separating peaceful exploration from military applications. Rather than tap one of the military programs to lead the U.S. effort, the president responded to calls for a U.S. response by founding a civilian organization: the National Aeronautics and Space Administration (NASA). By signing the National Aeronautics and Space Act in July 1958, he transformed the National Advisory Committee on Aviation (NACA) into a space agency with a name change, a new mandate, and expanded authority.[33]

Even after Eisenhower created NASA, however, the military services did not immediately give up their individual human spaceflight programs.[34] Private researchers also ran experiments without any concern that they were trespassing on NASA's exclusive domain. The failure to yield instantly and completely to the new civilian agency resulted from the way aerospace research had developed. Researchers conducting space-related experiments before *Sputnik* had grown accustomed to pursuing their interests despite supervisors' actively discouraging them. Those who persisted cared more about experimental results than bureaucratic boundaries. Even after NASA's founding, military and private aerospace scientists alike continued to pursue independent projects. Such disparate efforts did not threaten the new civilian space agency because in many ways it had been built to take advantage of existing programs. The field of space exploration remained relatively open in this period because NASA itself was adaptable.

NASA's founding legislation gave it the power to combine existing resources under civilian leadership. The agency's predecessor, NACA, a civilian organization charged with managing complex aeronautical testing and development since 1915, provided a stable basis to build on. When Eisenhower turned it into a space agency, however, NACA consisted of an administrative structure with only three major laboratories and two small ones. In order to proceed, the organization needed aerospace facilities. In its first three years, NACA turned NASA grew quickly by absorbing existing programs.[35]

NASA had the authority to draw such existing projects, including entire military space programs, into its purview. The navy's Project Vanguard satellite effort was the first external piece to be incorporated in this way. It joined NASA a month after the space agency's founding. Also in 1958, the California Institute of Technology's Jet Propulsion Laboratory (JPL) was reassigned to NASA control. Not all aerospace researchers welcomed the transfer. Despite army resistance to giving it up, in 1960 Werner von Braun's Army Ballistic Missile Agency, based in Huntsville, Alabama, at the Redstone Arsenal, became NASA's George C. Marshall Space Flight Center.[36]

In the first years after NASA's founding, this ability to absorb entire programs permitted significant organizational versatility. The space agency could expand into a new area of interest, or procure resources to fill a need, by annexing existing programs. Personnel, knowledge, equipment, and research would all be included. In this period a program's originating outside the space agency's auspices did not preclude its becoming an official NASA project at some point. In the years immediately after its founding, NASA's power to absorb programs underlay its success. In one significant area, however, the space agency chose not to designate an official site.

Although the space agency's need for medical services had been recognized from its very founding, NASA did not annex an existing facility, nor did it found its own. After many debates and a formal inquiry in 1959, the new agency chose not to develop its own medical research center. In 1962 NASA reaffirmed the decision, electing to continue using external facilities for astronaut testing. Its leaders deliberately chose to rely on external, independent aerospace medical facilities in order to avoid the expenditure of time and money involved in creating their own medical center. In addition, and no less significantly, this choice allowed NASA to avoid showing preference for any particular military service.[37]

Even after Eisenhower established the civilian space agency in 1958, the military services did not entirely give up their interest in piloted spaceflight. In the late 1950s and into the early 1960s, they maintained scaled-back versions of their human spaceflight programs. NASA's use of a particular military service's facilities for high-profile projects risked showing favoritism and thus reigniting interservice rivalries that had only just been forcibly doused.[38]

So in 1958, when NASA wanted to select candidates to become the United States' first men in space, it turned to the Lovelace Foundation for Medical Education and Research. The Foundation's civilian status gave it an advantage as the space age dawned. In addition, its other research projects already focused on areas important for extraterrestrial travel. As a consequence, the Albuquerque facilities had the diagnostic equipment and experienced staff needed to undertake a testing program on that scale. Because of the clinic's ongoing support of the U-2 pilots, NASA officials knew the Lovelace Foundation could handle the secrecy required for the initial testing stages. Lovelace himself also maintained a working relationship with the space agency. As head of NASA's Special Committee on Bioastronautics, he knew the physiological issues that concerned scientists considering human survival in space.

Beginning with Project Mercury, the Lovelace Foundation maintained a close connection with NASA but never became a part of the space agency. Likewise, though Lovelace himself held an official NASA title, he worked with the space agency as a contractor, not an employee. Despite its intimate role in selecting and monitoring the Project Mercury astronauts, the Lovelace Foundation never became a part of NASA at any point. Because the space agency chose not to create its own medical testing facility, however, the Foundation maintained a special association with NASA. As Dr. Kilgore aptly described the arrangement, "Lovelace was the *de facto* medical department of NASA because there wasn't any."[39]

Throughout their contract for the initial astronaut testing and in the years since, the Lovelace Foundation researchers took great pride in their NASA affiliation. Even physicians who worked on prestigious government contracts every day and had become accustomed to the top-secret U-2 project recognized that they were participating in something historic. The Lovelace researchers responded to the challenge of their new assignment with all their resources and experience. The development of the astronaut-testing regimen became their most famous achievement.

Creating the physical examinations used to help select the first American astronauts solidified the Foundation's reputation as a premier research center, making the Lovelace name represent the best in space medicine. The physicians organized a testing system that examined every physical condition the new American astronauts could possibly encounter in the void of space. Drawing on all their experience in high-altitude testing and their extrapolations about space conditions, the Lovelace researchers compiled a thorough and exacting regimen. Years later, Tom Wolfe made their methods famous when he described them in *The Right Stuff*. The depiction captured the examinations' trademark thoroughness, albeit somewhat unsympathetically (from the point of view of the poked and prodded astronaut candidates).[40] Although the Lovelace testing became the most famous segment of the Project Mercury astronaut selection, it was by no means the only component.

Before any physical testing began, NASA officials established a procedure for selecting the astronauts who would participate in Project Mercury, the space agency's missions to launch the first piloted suborbital and orbital flights. That procedure began with setting minimum requirements and narrowing the candidate pool. Although NASA planners initially considered recruiting civilian astronaut candidates, the inherent risks and classified nature of aspects of the space program soon made military personnel more attractive. NASA constructed the qualifications to balance age, education, and experience. All candidates had to be under forty years old and hold a college degree in engineering (or its equivalent). In addition, they must have graduated from military test pilot school. Finally, each needed at least 1,500 hours of flying time as a qualified jet pilot.[41] Requiring experience in test piloting jets significantly narrowed the candidate pool. That requirement came straight from the top.

Eisenhower's decision to choose the nation's first astronauts from among jet test pilots gave NASA a way to tap into a self-selected group of men who were trained to think like engineers and had already volunteered to risk their lives for their country.[42] As a group, these men flew the hottest jet airplanes the nation had. The natural selection of military test flying had winnowed the group to an elite corps. Through that flight experience, the jet pilots also mastered valuable skills that NASA wanted its astronauts to possess. Test pilots flew high-performance aircraft to detect problems, diagnose the causes, and clearly communicate the analyses to the engineers and mechanics. In addition, military pilots understood discipline, rank, and order. They could take orders. Selecting military jet test pilots as potential astronauts allowed NASA to choose from a cadre of highly motivated, technically skilled, and extremely disciplined pilots.

Selecting astronauts from this group also meant accepting other characteristics. The aviation idea of the "right stuff" that Wolfe adopted as the title and theme of his book about jet test pilots and astronauts summed it up. Having the right stuff meant possessing skill, daring, and an unwavering belief in one's own abilities. It meant working in an all-male environment where a certain coarseness complemented the dangers ever present in the work. It meant dealing with death. Having the right stuff meant exhibiting the particular brand of masculinity needed to strap oneself into an unproven aircraft for the express purpose of pushing that airplane to its limits. For some pilots with the right stuff, the ability to regularly risk one's life carried side effects: risk taking, fast driving, womanizing, and hard drinking, manifestations of a drive that could not be limited to the airfield. Such macho excesses did not worry NASA decision makers. The space agency viewed this particular kind of manhood as part and parcel of the talents NASA needed. And the space officials charged with overseeing the jet pilots turned astronauts already had plenty of experience with military aviation and all its ramifications.

The job of selecting the fittest of these men fell to the Lovelace Foundation. When the first group of pilots arrived in Albuquerque for testing, they had already survived several cuts. The military services selected pilots to report for astronaut testing based on their records and performance. Of 508 military pilots whose career files NASA reviewed, 110 satisfied the basic requirements. A closer evaluation of their records extracted 69 possible candidates.[43] After a briefing about the space program and a personal evaluation interview in Washington, DC, the group shrank again. Only 32 men went on to New Mexico.

Once there, the men began the Project Mercury physical: seven and a half days of closely scheduled tests. During that span, the Lovelace researchers, led by Dr. Schwichtenberg, conducted thirty different laboratory tests on each candidate. Eye and ear investigations examined not only sight and hearing but also balance and coordination. Thorough x-rays produced a complete picture of each candidate's entire body. To compile all the test results for each of the candidates, the physicians also used an innovative new method of record keeping: the computer. Researchers recorded their findings on punch cards that were run through an IBM machine to tally the data points. As expected, the profiles that emerged were impressive.[44]

Given the high level of physical fitness that each of these pilots exhibited in his daily work, the physicians did not expect to find disabilities or illness. Rather, the Lovelace researchers used established clinical baselines to search for unexpected weaknesses that might become problems under the stresses of a space flight. In the end, the physical examinations eliminated only one pilot. Rather than using the tests to weed out prospective astronauts, the Lovelace physicians compiled assessments that could be used to make recommendations to the final selection committee.[45]

Additional examinations at other sites further ranked the candidates. After Albuquerque, the astronaut candidates traveled to Wright Aerospace Medical Laboratory in Ohio for stress testing and psychological examinations. Finally, the NASA selection committee used the recommendations made by the Lovelace and Wright physicians to winnow the group down to seven astronauts. Originally the space agency had requested a final group of six, but because the selection committee could not make the last cut, they recommended seven men to NASA administrator T. Keith Glennan. Rather than eliminating one, Glennan accepted all seven.[46]

The astronaut examinations that the Lovelace Foundation organized for NASA in 1959 brought together the most advanced physiological tests of the day. The staff combined their research on human tolerance for spaceflight with the wealth of information they had gathered over the preceding years about survival in the aeropause. This time, however, the U.S. government and the military services openly supported the research. The derision that had kept aerospace medical experiments buried deep in aviation researchers' budgets evaporated when the space age dawned. With the potential for human spaceflight seeming more real every day, in 1960 Lovelace's curiosity about the wide-ranging possibilities for people in space led him to initiate the first systematic testing of women's capabilities for spaceflight.

The talent pool from which he drew female test subjects came into being at the same time when he and his colleagues began exploring the "upper atmosphere": during and after World War II. In this period, women's aviation opportunities underwent two radical changes. First, the end of the Women Airforce Service Pilots in 1944 severely limited women's professional aviation employment. Second, democratizing forces in aviation created ways for those without personal wealth to learn to fly. Despite the widespread availability of aviation in the postwar period, however, women found their opportunities constrained.

Even as aerospace medicine opened the upper atmosphere to men, the barriers blocking women from the vanguard of aviation slammed into place. Bans on women's participation in military flying systematically excluded them from the testing ground for the newest aviation technology. In addition, aerospace medicine's prohibitive costs and exclusive access made it hard for women to demonstrate that they could handle the work. Within the severe limits placed on their aviation opportunities, however, female pilots created organizations and competitions that encouraged women to fly. By the late 1950s, the small but thriving world of women's aviation produced a significant pool of talented pilots from which Lovelace drew the candidates for his Woman in Space Program.

3

WASPS, WHIRLY-GIRLS, AND NINETY-NINES

Female Pilots and Postwar Women's Aviation

On August 4, 1944, Jean Hixson marched across the training grounds at Avenger Field in Sweetwater, Texas, in formation with her seventy-two classmates, all graduating members of the Women Airforce Service Pilots (WASP), class 44-6 (the sixth class of 1944). Through months of ground school, flight instruction, and military exercises, she and her classmates had endured sixteen-hour days of training to serve their country as nonmilitarized pilots during wartime. With their graduation, they stood ready to assume stateside aviation duties, freeing male pilots for combat roles in the European and Pacific theaters. Even as Hixson received her WASP wings, however, Congress, the media, and the military fiercely debated the fate of the WASP program. After the ceremonies, Hixson and the other 44-6 graduates received flight assignments even though the program's future remained uncertain. But on October 1, 1944, the War Department finally announced its decision: the WASP program would end on December 20, 1944—while the war still raged. Henceforth the air forces banned women from flying military aircraft.[1]

After the WASP disbanded, Hixson returned to her Ohio teaching job. She spent the rest of her aviation career in the Air Force Reserve. In 1957, as part of a study outline she was making for the Akron school system, she flew through the sound barrier as a passenger on a jet, one of only a handful of women to accomplish that feat. A year later, she took the controls of a Delta Dagger, another jet, while it was in the air. Because regulations forbade women from piloting military aircraft, a male pilot took off and landed the airplane, but Hixson enjoyed the rare experience of handling the jet during its flight. Because the post-WASP ban on women flying military aircraft remained in effect for years after World War II, neither of the two highlights of Hixson's Reserve years involved her piloting the advanced aircraft alone.[2]

Hixson's flying career encompassed both the uncommon expansion of women's opportunities during World War II and the forcible restrictions women endured in the postwar period. Although few of her fellow women pilots maintained the military connections that Hixson did (she retired from the Air Force Reserve as a full colonel in 1981), many female pilots found ways to fly in the decade and a half after the war. Despite being barred from flying the military's most advanced aircraft for the bulk of her service, Hixson nurtured her passion for flying. She did so by taking advantage of established support systems. During the 1950s, female pilots combined the resources created by the wealthy female pilots of the 1920s and 1930s with the public aviation courses that made flying accessible to the middle and working classes.[3]

Without an examination of this historical context, the women who took Lovelace's tests (as Hixson did) seem extraordinary—truly exceptional. In fact, hundreds of women flew airplanes in the 1950s. When Lovelace and his colleagues began recruiting female pilots to take his Foundation's astronaut tests, the women he invited represented a cross section of aviation's female elite. Taken collectively, their experiences reveal the strong networks that maintained flying opportunities for women in the 1950s. They participated in organizations founded to promote women's aviation by connecting women pilots to each other. Together these dedicated women worked within the limits of the military ban to do what all pilots wanted to do: take advantage of opportunities to challenge themselves in the sky. When the space age dawned, Lovelace's Woman in Space Program became possible, in no small part, because female pilots had fostered a small but thriving world of postwar women's aviation.

Consequences of WASP Disbandment

During and after World War II, aviation ceased to be the domain of the rich. For both men and women, flying became more accessible to middle- and working-class people. For men the change occurred during the war when the Army Air Forces trained legions of pilots. Afterward, the GI Bill provided returning veterans with tuition grants that could be applied to academic aviation programs. But unlike their male counterparts, for whom the war opened flying jobs that continued after peace was achieved, women pilots found that aviation opportunities expanded only briefly during World War II and narrowed considerably even as active combat still continued. For women, professional flying options never fully materialized.

Before and during World War II, two programs intended to supplement men's military flying gave women access to more affordable aviation lessons. One, the Civilian Pilot Training Program (CPTP), opened the field to novices, while the second—a set of two programs eventually consolidated into the WASP—ushered women into more advanced flying. Both would eventually be curtailed in order to concentrate available resources on male pilots.

The Civilian Pilot Training Program made flying more accessible to middle-class women. In 1939 Congress approved a New Deal bill to benefit private aviation and promote preparedness by dedicating federal money to producing licensed pilots. The program funded aviation classes at colleges and universities. Organizers assumed that women's control of the household purse strings meant they would have to support aviation—or even be able to pilot the family airplane—in order for private aviation to flourish. From its inception, therefore, the CPTP stipulated that for every ten men, one woman could participate. Congress's antidiscrimination provision opened the CPTP to African American men and women. Several women's colleges also established CPTP programs, greatly increasing the number of female licensed pilots.[4]

As World War II loomed, however, preparedness became the CPTP's primary goal. The program focused on training fliers who could join the air forces when the United States entered the European war. By 1941, new regulations required that all trainees, male and female, sign a pledge promising to join the armed services if they were needed. Since women could not serve in combat, the CPTP officially barred women beginning in June 1941. In the meantime, however, the program allowed many women to earn pilot's licenses. After the United States entered World War II, the existence of so many trained female pilots inspired plans to utilize them for the war effort.[5]

Similar programs to employ women's talents in aid of the various armed services emerged throughout 1942. The navy created a women's reserve called the WAVES (Women Activated for Voluntary Emergency Service). In the coast guard, the women's reserve took its name, the SPARS, from the service motto, *Semper paratus*—always ready. The army initially created the Women's Auxiliary Army Corps (WAAC) to work within the U.S. Army but not be a part of it.

Later the army renamed the militarized group the Women's Army Corps (WAC). The idea for a women's auxiliary of the Army Air Forces emerged before the United States entered the war. After Pearl Harbor, however, two women proposed different ways of tapping the pool of female pilots.[6]

The differing proposals for using women's flying talents became the Women Auxiliary Ferry Squadron (WAFS) and the Women's Ferrying Training Detachment (WFTD), later to be united as the WASP. In 1941, after Jacqueline Cochran ferried a bomber across the Atlantic for the British Air Transport Auxiliary (ATA), an ATA officer asked her about training American women pilots to help the ATA. Cochran responded by recruiting twenty-four women, training them in Canada, and traveling with them to England to enlist. Before she left, Cochran elicited a promise from Gen. Hap Arnold, the head of the Army Air Forces, that no women's program would be started in the United States without her.[7]

At the same time, Nancy Harkness Love, an accomplished pilot who was married to Robert Love, the deputy chief of staff for the Ferry Division of the Air Transport Command (ATC), got an ATC job and began planning a women's service. In September 1942 the WAFS began to organize women pilots to ferry military aircraft. By the time Cochran returned from England that fall with General Arnold's approval to head the WFTD, a project designed to teach less experienced female pilots to fly military aircraft, the WAFS had already begun. For a time the two women's programs coexisted—organized in different parts of the AAF and serving different purposes. Less than a year later, however, the two programs united as the Women Airforce Service Pilots, with Cochran as the overall director of women pilots and Love remaining as the head of the WAFS under Cochran.[8]

As with the Collier Trophy, Cochran's campaigning tactics and personal contacts within government channels facilitated the WASP's consolidation. Cochran's staunch advocacy for her interests, and her willingness to oppose those who did not see things her way, fit a pattern of behavior that characterized her public actions throughout her life. When a cause had been lost, however, she exhibited a pragmatic willingness to move on to other projects.

The progression that led to the disbandment of the WASP began as a result of conflicts over militarizing the civilian organization. After Cochran refused to allow the WASP to join the WAC, a congressman proposed a bill to militarize the WASP as a part of the Army Air Forces, with full benefits. Male pilots lobbied actively against this change, however, fearing that deploying women for stateside duties meant male pilots would be reassigned to ground forces. After the bill was defeated, Congress directed the AAF to discontinue further WASP training. Existing WASPs ended their service on December 20, 1944; the Army Air Forces did not rehire them. Although Cochran had worked hard to maintain the WASP, she also assessed the program's future pragmatically. Rather than continue a futile fight, she set her sights on new accomplishments. After the WASP program ended, Cochran remained the only woman in the AAF.[9]

Individual WASPs scrambled to find other ways to serve. On October 7, two days after Arnold announced the coming disbandment, Jean Hixson wrote to the Chinese embassy in Washington, DC, volunteering to fly for the Chinese military. The reply rebuffing her "kind offer" did not even recognize that she was female. In a letter addressed to "Dear Sir," the Chinese attaché suggested she apply to the United States War Department. She tore the response in half. Hixson could not use her piloting skills to help the war effort.[10]

Other WASPs also felt frustrated by the unceremonious dismissal. When the final December 20 deadline came, the U.S. Army provided only a military transport back to the airbase nearest each WASP's home, at her own expense. Even then, not every commander supplied the required transportation. The hurried push to return to peacetime normality denied the pilots any recognition for their sacrifices and service. Thirty-four years later, in 1977, Congress finally militarized the WASP, allowing the surviving participants to collect veterans' benefits.[11]

In the intervening years, however, the WASP's end significantly limited women's postwar aviation. After 1944, the armed forces banned women from flying military aircraft. Because the military tested most new aviation technology, women lost their access to the forefront of the field. Military test flying programs "wrung out" the newest aircraft before manufacturers created tamer versions for eventual public sale. Imposing the ban on female pilots at the same time that jet aircraft came into use made the effect doubly damning. As aviation increasingly relied on the jet engine, women lost the means to obtain advanced flying experience.

Very few women found ways to access jet technology. During World War II Ann Carl, a WASP stationed at Wright Patterson Air Force Base, became the first woman to fly a jet. At the time, jet technology remained a state secret, requiring that the parked aircraft's jet propulsion be disguised by a fake propeller. If not for the military ban on female pilots, a few more women might have been able to maintain their positions in flight testing and other elite niches of aviation. As it was, no American woman flew another jet until Jackie Cochran broke the sound barrier in a Sabrejet F-86 in 1954. Cochran's access to jet technology came in spite of her gender, however, and because of her money and connections. Her husband, Floyd Odlum, used his substantial financial interest in the Canadian manufacturer of the Sabrejet to request that Cochran be allowed to test the aircraft.[12]

As male pilots returned from wartime service to take aviation jobs, even female pilots with advanced credentials had difficulty finding commercial jobs. Major airlines echoed the military's ban on women pilots. After World War II, both military and commercial professional aviation became largely closed to women. Jobs remained available in general aviation: working for small freight carriers, teaching flying lessons, chauffeuring executives, and dusting crops, but none of these options offered the security of standard commercial aviation. The prohibition against female military pilots that resulted from the WASP's disbandment fundamentally limited aviation opportunities for women until the 1970s.[13]

Postwar Women's Aviation

At the same time that women's commercial and military aviation opportunities shrank dramatically, however, the chances for a young woman to pilot an airplane expanded. Through colleges and civic groups, learning to fly became widely available for middle-class youth in the immediate postwar era. Wartime pilots became peacetime flight instructors who needed students. College competitions and the Civil Air Patrol provided places for young women interested in aviation to practice and improve. Many of the women who would later be invited to take part in Lovelace's astronaut fitness testing entered aviation through these avenues. Although the number of women pilots remained small during the 1950s, the cohort grew, and the networks they maintained allowed women's aviation to flourish.

Many women pilots in the 1950s formed their opinions about women's capabilities during World War II. As Jerri Sloan Truhill remembered, "At that time [during the war], women were ferrying airplanes. Women were doing all sorts of jobs that men had been doing. They were running trains, for goodness sake. They were driving buses. Women were working in factories building airplanes." Women's wartime opportunities inspired her. She saw that women could succeed in nontraditional jobs, and "just because the war was over we didn't know that they couldn't do them anymore." Women who yearned to fly saw the WASP program as proof that their dreams could be realized. Irene Leverton, Jane B. Hart, and Myrtle Thompson Cagle, three women who later took Lovelace's tests, all learned to fly during World War II with thoughts of joining the WASP.[14]

For two of these women, such opportunities would not have been possible twenty years earlier. Myrtle Thompson's family was not well-to-do, and Irene Leverton's parents were solidly working-class. Of the women pilots who later participated in Lovelace's Woman in Space Program, only Jane Hart could be described as a wealthy woman pilot. As the daughter of a Michigan automobile manufacturer who owned the Detroit Tigers baseball team, she earned her private pilot's license despite her family's aversion to flying.[15] The modest financial backgrounds of the rest of the women contrast with the well-to-do profiles of women fliers from the 1920s and 1930s, however. Most categorized their families as working-class or middle-class.

They exemplified the many middle-class and working-class women who found ways to pay for flying lessons in the postwar period. A young woman willing to sacrifice her "fun money" and spare time could accumulate hours and earn ratings steadily. Sarah Gorelick Ratley managed her finances so that she could maintain an airplane and keep flying. She reasoned that "you always have the money for the things that mean the most to you. You would live in a less expensive apartment, not have a car, ride the bus, [or] not buy as many clothes." By managing her money with aviation as her top priority, she found the funds to keep flying.[16]

The widespread availability of aircraft and qualified instructors after World War II made flight instruction more accessible. Returning male veteran pilots and abandoned WASPs used their flight instructor ratings to offer lessons and establish aviation schools. Postwar flight schools became revolving doors of returning pilots looking to start civilian aviation careers. The competition for clients meant that instructors needed to set reasonable rates to attract students.

Compassionate flight instructors sometimes also allowed students to work out creative methods for financing lessons. Washing and waxing aircraft could be exchanged for time aloft. Virginia Holmes made a deal with Mrs. Browning of Browning Aviation in Austin, Texas, to maintain the pilots' logbooks in return for lessons. Without the funds to sustain their expensive hobby, aspiring pilots had to be both frugal and innovative. Bernice "B" Trimble Steadman's instructor insisted that students fly at least twice a week so they would not forget too much between lessons. Since she could not afford more than an hour a week, she met his requirement by flying twice a week, thirty minutes at a time. After she earned her commercial license, she began working for the same flight school where she was taking her lessons.[17]

Friends or siblings often drew women into aviation. One of Steadman's closest friends began taking flying lessons soon after she did because she feared they would never get to see each other unless she also learned to fly. They both earned private pilot's licenses. Another women who later took Lovelace's astronaut examinations, Gene Nora Stumbough Jessen, learned to fly because she followed her brother's interests into the Civil Air Patrol (CAP). When her older brother wanted to join the flying civil defense organization but could not attend the meetings because of his evening job, she went to the gathering with his girlfriend, Elaine. They both signed up. In the postwar period, the Civil Air Patrol opened the door for many women, men, and adolescents to begin flight training in the aim of civil defense.[18]

The Civil Air Patrol was organized in late 1941 to mobilize civilian pilots for stateside duties in the impending war. When the United States entered World War II, the CAP flew coastal patrols looking for German submarines, protecting oil tankers, and guarding against another strike like Pearl Harbor. In the first few weeks, CAP volunteer fliers rescued a disabled tanker off the coast near Atlantic City and disrupted a submarine attack on a tanker near Cape May, New Jersey. Within months of its founding, the CAP numbered 40,000 volunteers. From its beginning, the group encouraged discipline by using military-style uniform insignia, airplane markings, and marching drills. In 1943 the CAP became an official auxiliary of the Army Air Forces. After the war, it continued its mission of national defense by maintaining a well-prepared, flight-ready citizenry.[19]

The Civil Air Patrol drew recruits from all sectors of American society. Adolescents could join the CAP Cadets, the youth training branch. African Americans also participated actively in the CAP. Willa Brown, an accomplished African American pilot in the 1930s, served as a CAP officer throughout World War II. During and after the war, the CAP actively recruited female members. By 1945, women made up 20 percent of the members. With the exception of not permitting female pilots to fly coastal patrols, the CAP allowed women to take part fully in both teaching and receiving flight training. In the decade following the war, women maintained a significant presence in the civil defense association. By 1956, 14,000 female members accounted for over 15 percent of the CAP's total of 91,000 participants.[20]

The Civil Air Patrol put flying lessons within the reach of any young person, regardless of means. For example, Irene Leverton would never have learned to fly without the CAP. Her mother, a scrubwoman and the widow of a Chicago ironworker, scraped together the money for a uniform so that Leverton could take flight training with the Chicago chapter in 1944. Leverton walked five miles between the bus stop and the airport for a half-hour of flying every two weeks. She used her flight training to begin a long career as a commercially licensed pilot.[21]

Civil Air Patrol members could take advantage of group lessons to get flight training and airtime. The CAP also offered ground school, the classroom education required to receive aviation ratings (certifications of advanced skills). Some squadrons found ways to extend their flying instruction beyond the official training exercises. Georgiana T. McConnell remembered that members of the Nashville chapter formed a club to buy a Piper Cub and took turns flying it to accumulate hours and experience.[22]

For many female pilots, the Civil Air Patrol provided a stepping-stone to the other forms of aviation training. Gene Nora Stumbough Jessen recalled in particular, "I came from an extremely modest home, never would have been able to afford to fly airplanes, never entered my mind that I could fly airplanes." After taking initial lessons with the CAP, she enrolled at the University of Oklahoma where, inspired by her CAP experience, she took aviation classes. The costs of flying challenged students of modest means, but they could be overcome. After her parents paid for her to earn her private pilot's license during her sophomore year, Jessen remembered, "I couldn't do both: aviation and college. I just didn't have that kind of money. . . . I would drop out a semester. I would go to work [flying]. And then I would come back and I would borrow a little money." Eventually she became a part-time flight instructor at the university so that she could continue to take classes. After six years she graduated with a degree in English and a résumé as a flight instructor. Her experience of developing aviation skills and earning ratings while in college exemplified another way that many women without personal wealth gained aviation training during the 1950s.[23]

In the 1940s and 1950s, many high schools and colleges added aviation classes to their curricula. After the United States entered World War II, Georgiana McConnell took the aviation course offered at her high school during her senior year. Classes like hers exposed young people—both male and female—to the allure of flying. Many colleges also offered aviation as a field of study. On-campus programs often offered both basic flying lessons and ways to earn advanced ratings.

In 1941 Stephens College, a Columbia, Missouri, women's college, took the idea of aviation classes a step further than the CPTP and other programs by innovating an entire curriculum based on "air-mindedness." Students prepared for the coming aerial age by learning about passenger service, aviation administration, and aerial navigation. The school also offered a piloting option. Marilyn Link, whose brother Edwin invented the Link Trainer (the first airplane simulator), followed her passion for aviation to Stephens College, earning her private pilot's license in 1945 as the first step in a long aviation career. Although not all schools offered as extensive a curriculum as Stephens College did, in the 1950s many universities offered not only flying courses but also flying competitions.[24]

Intercollegiate aviation meets offered young pilots a way to test their new skills and freshly earned licenses against pilots from other campuses. Just as for other intercollegiate sports, schools fielded flying teams. The National Intercollegiate Flying Association regulated such competitions. Judges presented individual and team awards in a variety of events. Women competed regularly. In the late 1950s, Wally Funk got her mother's permission to take flying lessons after she encountered aviation classes at Stephens College. When she earned her private pilot's license, she followed her passion to Oklahoma State University, where she became the Flying Aggie top female pilot for two years running. Like many other female pilots, Funk developed her skills in collegiate meets.[25]

Even as more women became pilots, however, women remained a significant minority in aviation. The sex ratios were often extreme. Pat Jetton used her Air Force Reserve service to pay for flying lessons in 1948. As she recalled, "The school was one of the largest in the area and mostly male. For a long time I was the only female among almost three hundred students who were attending on the World War II GI bill." Women stood out in these crowds, and were forced to prove themselves continually. Such skewed sex ratios presented another challenge that female pilots needed to meet.[26]

Since female pilots remained the exception in aviation schooling, flying teams, and ground school classes in the 1950s, any woman who wanted to succeed as a pilot had to find a way to cope with her conspicuousness. The female pilots who thrived in the male world of aviation did so because they were talented, determined, accomplished, and flexible. As the only woman on her college flying squad, Gene Nora Stumbough Jessen put up with her teammates' nominating her (against her will) for the "Sky Queen" flying pageant because she was friends with the men in the group. She was more interested in winning legitimate flying awards than a combined beauty pageant and aviation meet. But Sarah Gorelick Ratley found that being the only woman in the engineering classes she took at the University of Denver beginning in 1951 could be fun. She got along well with her classmates, relying on them to help keep her eleven-year-old Ford running.[27]

As a group, the female pilots of the 1950s were independent women who sought out support for their ambitions. They did not aim to flout social conventions; they just wanted to fly airplanes. Increasingly, they came from middle-class families who sacrificed to send them to college. Rather than relying on personal wealth or sponsorships as 1930s "society-girl pilots" had done, postwar women pilots worked for wages or barter to keep themselves flying. They also relied heavily on woman-centered flying organizations.

Widening Their Worlds

In the postwar world of aviation, women's flying organizations connected female pilots with each other and encouraged more women to fly. The need for such associations became especially apparent as women noticed their modest population. Licensed female pilots remained a growing but still relatively small group. Inspired to connect with other women with a passion for flying, female pilots created new organizations to promote their interests. At the same time, the women's aviation clubs originally founded by well-to-do pilots in the 1930s expanded as more women entered aviation. During the postwar period, social organizations, air meets, and transcontinental races created a thriving world of women's aviation.

In the mid-1950s, Jean Ross Howard founded an organization for female helicopter pilots. When Howard earned her helicopter pilot's license, she learned that only twelve other women in the world had acquired helicopter training before she did. She invited them to meet in Washington, DC, to discuss their common interests. Together they founded the Whirly-Girls, the International Women Helicopter Pilots, on April 28, 1955. The whimsical name expressed the joy they found in flying helicopters. Starting with Hannah Reitsch, the Nazi test pilot who was the first woman to fly a helicopter, each of the Whirly-Girls took a number based on when she earned her helicopter rating.

They elected Howard, number 13, to handle the new club's business; she served as the Whirly-Girls' executive director until 1988.[28]

The numbering system the Whirly-Girls used combated the sense of isolation that female helicopter pilots felt when they examined their flying circumstances. When Howard founded the group, the thirteen members lived scattered all over the world. As new members applied, each of the new helicopter pilots continued to receive a number. By 1961 the organization counted forty-one female helicopter-rated pilots. Having an assigned number reinforced the awareness that despite feeling alone in a male-dominated area, each Whirly-Girl belonged to a group of women who shared her interests. Decades later, Howard questioned the practice of numbering each Whirly-Girl when the group's membership grew large enough that she wondered whether the tradition had lost its significance. The members would not hear of dropping it. The sense of belonging that the membership numbers symbolized remained one of the key benefits the Whirly-Girls offered.[29]

From its founding, the Whirly-Girls provided camaraderie, networking, and support. The founding meeting in April 1955 became the first Hovering. Each year after that, the Whirly-Girls organized another Hovering, a gathering of female helicopter pilots, to foster sisterhood among its members. The annual Hoverings provided a place for the women to share experiences. The organization also offered some extraordinary opportunities. In 1961 Jane Hart, number 25, arranged for the Whirly-Girls to meet President John F. Kennedy in the White House.[30] At a time when even fewer women flew helicopters than flew fixed-wing aircraft, the Whirly-Girls served a vital purpose by connecting women helicopter pilots to each other.

Many of the Whirly-Girls also belonged to another women's flying organization that survived from the 1930s: the Ninety-Nines. Female aviation pioneers Amelia Earhart and Ruth Nichols had begun planning the women's aviation organization two years before the 1929 Women's Air Derby provided an opportunity to found it. Although the derby, the first all-female cross-country competition, became known as the "Powder Puff Derby," the event's importance did not escape the participants. The women who flew in the derby came away from the race impressed with the need for a formal organization for women in aviation. Three months later, twenty-six women finally met at Curtiss Field on Long Island, New York, to organize the group. They sent invitations to all 126 licensed female pilots in the country: 99 accepted. At Earhart's suggestion, the new organization took its name from the number of its founding members.

The Ninety-Nines promoted women's aviation by pursuing charity work (dropping leaflets to publicize the Salvation Army's unemployment relief fund) and advocating aerial markings (painting identifying marks on public buildings and landmarks to help pilots navigate). In addition to fostering fellowship and promoting charitable causes, the group provided moral support for its members, who flew despite facing significant discrimination. The Ninety-Nines aimed to advance both aviation and the women in it.[31]

The international organization of women pilots that Amelia Earhart and ninety-eight other female pilots founded in 1929 thrived in the 1950s because it provided a place for women pilots to meet, socialize, and support each other. After World War II, enough professional women pilots existed for the Ninety-Nines to have become a professional organization if the group had chosen to go in that direction. Instead, the Ninety-Nines remained focused on promoting the personal connections between female pilots. This "social club format" fostered a sense of community by providing a support structure for women's flying.

Through the organization, members met fellow pilots. Many lasting friendships began in the Ninety-Nines. Janey Hart and Bernice "B" Steadman first met at a Ninety-Nines meeting in Flint, Michigan. When Hart wanted to earn her instrument rating, Steadman became her instructor. In addition to fostering such one-to-one relationships, regional chapters held regular meetings, organized charitable drives, and hosted fly-in gatherings.[32]

Each month, the Ninety-Nines' national newsletter publicized its members' accomplishments. Cover stories reported on the group's annual conventions and other significant women's aviation events. Members could also keep up with individual Ninety-Nines through detailed chapter reports. The regional chapters promoted women's flying by hosting regular meetings, aviation events, and even small competitions. Reporters from each region highlighted their group's activities in regular columns printed in the national newsletter. The attendance or addition of a new "49½er" (a Ninety-Nines husband) always deserved note. The regional chapters also routinely recognized their members by name for new flying achievements. Many of the kudos came from air races and competitions.

After World War II, the Ninety-Nines worked to restart women's air racing. Former WASPs played a significant role. In 1947 Mardo Crane, a former WASP, and the Florida Ninety-Nines invited West Coast chapters to attend their Tampa All Women's Air Show, featuring a number of accomplished former WASPs. Although this first women's transcontinental air race had only two contestants, many pilots attended the air show. Later that same year, the Ninety-Nines sponsored a very well attended women's aviation convention in San Antonio, Texas.

In 1949 the Ninety-Nines revamped the cross-country race, calling it the All-Woman Transcontinental Air Races (AWTAR), although public commentators revived the patronizing moniker first used in 1929: the Powder Puff Derby. National attention took off. The Ninety-Nines established an official set of rules and formalized timekeeping methods. Eventually the logistical complications of the annual event required the establishment of its own office. For the women who participated, the teamwork between pilot and copilot combined with the appeal of friendly competition to attract entrants. One woman in particular demonstrated that she was a superb competitor. From 1951 to 1969, Fran Bera flew in nineteen consecutive ATWARs, placing in the top ten on seventeen of those occasions and posting seven wins. She was one of the many highly accomplished women pilots who demonstrated her skills in these races.[33]

In addition to the AWTAR, the Ninety-Nines' regional chapters organized and promoted smaller women's air events. Beginning in 1947, the Illinois Ninety-Nines ran an annual air meet in Chicago. Women competed for awards in a variety of events that allowed them to demonstrate their skills. Irene Leverton, the organizer of the 1949 and 1950 Illinois Ninety-Nines air meets, recalled that such events not only allowed Chicago contestants to fly in front of home crowds, but also publicized the accomplishments of visiting pilots: "The gals from out of town, their home papers had it. It was a way of putting women pilots in front of the public." Other chapters also sponsored major events. In 1957 a Ninety-Nines chapter in Texas organized the Dallas Doll Derby, a three-hundred-mile all-women air race. Within five years, the race became the second largest all-women race in the country, second only to the AWTAR. Jerri Sloan Truhill won the Dallas air championship three years straight, from 1959 to 1961. When she remembered her participation, however, she took pains to note that events such as the all-women races did more than just publicize women's aviation achievements and promote healthy competition.[34]

Flying air races allowed female pilots to accumulate hours toward a commercial license, without which a pilot could not sell her services. As Truhill explained it, "The gals all raced, number one, because it was fun. And number two was to build up flying time because that was the only way that you could do it. Unless you just went boring holes in the sky." In order to advance, and to start making money by flying, pilots needed a way to get the experience required for advanced ratings. Transcontinental air races, record-setting attempts, and competitive flying in meets (whether club sponsored or intercollegiate) allowed pilots to accumulate enough flight time to turn their love of flying into a paying job in aviation.[35]

In the 1950s, many women pilots found work teaching flying, dusting crops, and piloting air taxis. Some companies hired female pilots to chauffeur their executives. Marilyn Link, later a teacher and pilot for the Nebraska Department of Aeronautics, first worked as a corporate charter pilot. Even with advanced credentials, however, many female pilots found that getting a foot in the door was challenging.[36]

Getting the first aviation job often required great patience and a lot of time doing other nonflying jobs. Jerrie Cobb's story illustrated the roadblock that many women pilots faced: employers did not expect pilots to be female. In 1951 she responded to an advertisement looking for DC-3 pilots willing to work without pay, for experience only. Cobb forwarded her qualifications, then drove from Oklahoma to Miami in response to the employer's interest. When she arrived, however, the firm had no job for a "girl pilot." She applied for an apprentice mechanic position at the Miami International Airport, but airport officials offered her a typist/file clerk position instead. Finally, she got her first flying job when she volunteered to help a stranded pilot with aircraft he was ferrying to Peru.[37]

Sometimes women's aviation events provided a direct connection to a flying job. In the late 1950s the Ninety-Nines' Illinois air meets inspired Chicago city officials to invite the organization to dedicate Meigs Field, the lakefront airport. During the ceremonies, Irene Leverton (who supported herself as a Rand-McNally artist) flew a stunt in which she "dusted" the crowd with a donated gallon of Jacqueline Cochran's "Tailspin" perfume. (On Leverton's first pass, the wind blew the wrong way. Instead of spraying the spectators, she doused about 150 Chicago policemen on the other side of the runway. "I said, that is the first time Chicago cops smelled good. But anyway, then I made a pass in the back of the stands and beat it.") In the stands, her friend sat next to the head of a western Illinois aviation company. When he commented on the stunt, the friend told him that Leverton was an old-time "ag" pilot, which she was not. The firm hired her.[38] Leverton used the position to earn her instrument rating (and used her pay to buy her mother a used car and a refrigerator). Being able to earn money in aviation allowed her to continue to fly, accumulate ratings, and make herself more competitive as a working woman pilot.

Women also used aviation to help them become partners or owners in business enterprises. In 1952 Myrtle Thompson Cagle opened a one-woman flying operation at the Selma, North Carolina, airport. She administered the business, taught lessons, and repaired aircraft. She also wrote an aviation column for the *Raleigh News and Observer.* Jerri Sloan and Joe Truhill, the man who later became her husband, created a partnership to do test flying for Texas Instruments. When they won the contract, they formalized Air Services, a testing company. Another Ninety-Nine founded her own flying school. Bernice "B" Trimble Steadman learned to fly during World War II. In order to be competitive as a flight instructor after the war, she earned all her ground instructor ratings (required to teach the classroom component of flying lessons). At a student's suggestion, she struck out on her own to found Trimble Aviation—her own flying school—with a borrowed office, some military surplus furniture, and fourteen students. Despite the limitations on women's professional flying in the postwar period, a small but significant group of women established themselves in private flying businesses and other aviation careers.[39]

For those who did not pursue aviation full time, recreational flying provided many women pilots with a ticket to the wider world. During cross-country flights and in races, they saw parts of the country that they never would have been able to visit otherwise. Several air races traversed the nation each year with stops in major cities and gala celebrations at the end points. Female pilots even got to visit other countries. Less than a year after Jerrie Cobb's long struggle to get her first flying job led her to Lima, Peru, she landed another international aviation gig. Just before Christmas in 1953, she flew a B-17 to Paris, France. As it had for many women pilots in the postwar period, her pilot's license opened the world to her. Women's aviation organizations worked to ensure that all female pilots shared the same chance.[40]

In the summer of 1960, the Ninety-Nines sponsored a grand tour of Europe. As an "international organization of women pilots," the group planned the trip to strengthen ties between the American chapters and their international counterparts. Indeed, after the Ninety-Nines' president returned to the United States, she reflected in the official newsletter that the trip offered "opportunities . . . to become better acquainted with our sister pilots across the ocean and [gain] a better understanding of their flying problems." For many of the Ninety-Nines rank and file, the grand tour was another way aviation widened their world.[41]

Fifty-four women rearranged their schedules and planned their vacations so they could spend part of the summer traveling through five European countries. Sarah Gorelick Ratley, an engineer working at AT&T and Long Lines Engineering in Kansas City, saved her money and took extra vacation time in order to make the trip. Georgiana McConnell quit her twelve-year job as a life insurance agency cashier to go on the tour. Ninety-Nines from all over the country packed their bags and joined the commercial transatlantic flight.[42]

For the next few weeks, special aviation events highlighted the sightseeing trip across Europe. In Great Britain, the Ninety-Nines began their day with a walking tour of London, then ended the evening with a reception at the Royal Aero Club. Visits to Oxford and Stratford-on-Avon followed. The Ninety-Nines organizing the trip also planned appointments for the visiting American pilots to test out European air space. As Ratley recalled, "We were flying aircraft in England, and Holland—throughout the Continent." Dutch pilots treated the Americans to an afternoon in light planes and gliders. In Belgium, with fewer airplanes available—not enough for the entire group—twelve lucky travelers had a chance to take to the sky once more. The tours of Germany and Italy also featured aviation highlights.[43]

In addition to these special aviation-related events, the travelers experienced all the adventures that arise from being new visitors to a foreign land. Some of their luggage went astray temporarily. As they moved from country to country, the women continually readjusted to which side of the street automobiles used and therefore which direction they needed to look before stepping off the curb. They also faced each country's idiosyncrasies. In London, Ninety-Nines looking for a bathroom break ended up digging frantically through their purses when they discovered that they needed an English copper penny to pay for public restrooms called "penny stops." In Amsterdam, tour members who mistook Dutch showerheads for similarly hung French telephones ended up with an earful of water. Back in the United States, the participants recounted these episodes to great effect at regional Ninety-Nines gatherings. One member of the tour wrote them up for publication in the national newsletter.[44]

An event like the Ninety-Nines grand European tour illustrated the opportunities that many working-class and middle-class women found in postwar aviation. Despite the severely curtailed access to commercial flying that resulted from the WASP's disbandment, women's aviation thrived during the 1950s. Women honed their skills through CAP training, intercollegiate competitions, and women's air races. They pushed themselves to perform, seeking out exciting new opportunities. By the late 1950s, a group of talented, well-credentialed female pilots stood poised to take on new challenges. Several of them joined the Ninety-Nines European trip.

The organizers of the 1960 grand tour hoped the excursion would allow American Ninety-Nines to interact with their European counterparts. And the pilot tourists did spend a lot of their trip talking to Continental Ninety-Nines and other foreign aviation aficionados. But the group of women fliers talked to each other as well. The extended trip let the female pilots get to know each other quite well, as happens so easily when traveling in close quarters through a foreign land. But not all of the talk on the buses and in the hotels centered on what they had seen in their travels: the European flying facilities, British landmarks, German beerhouses, or Italian architecture. As the Ninety-Nines traveled through Europe in the summer of 1960, rumors flew through the group. Back in the United States someone was interested in testing women pilots as potential astronauts.[45]

4

Betty Skelton, Ruth Nichols, and Jerrie Cobb

The rumors were true. Back in the United States, several prominent aerospace researchers, including Dr. William Randolph Lovelace II, had been evaluating women as potential astronauts. The idea of women as space travelers was not new. By 1959, various scientific studies suggested that women possessed traits that suited them for spaceflight. But the research supporting the idea had been sporadic. The time was ripe for a more serious investigation.

In fact, three plans for testing women emerged almost simultaneously in 1959 when—within weeks of each other—a trio of aeromedical organizations explored women's potential for spaceflight. First, *Look* magazine used NASA's own facilities to put Betty Skelton, a famous aeronautics champion, through astronaut exercises. Second, about the same time, in October or November 1959, air force researchers began aerospace tests on aviation pioneer Ruth Nichols. Third, Jerrie Cobb took the Project Mercury physical at the Lovelace Clinic in Albuquerque, New Mexico. Although Cobb took the Lovelace examinations in February 1960, the plan that led to her testing—an air force proposal called Project WISE (Woman in Space Earliest)—had originated five months earlier, about the same time as the two other women's testing projects.

In many ways, 1959 was the right time to explore the idea of women in space. Several researchers shared the idea. And at the beginning of the space race—before any person had flown in space—it seemed as if women's inclusion might be possible. And yet the time also presented many obstacles to considering women as serious space candidates. Of the three experiments initiated in the autumn of 1959, only Cobb's testing represented more than covert curiosity or a publicity stunt.

At this historical moment, Lovelace's Woman in Space Program began because a delicate balance of personalities, institutions, and ideals brought it into existence. As the project developed, however, the momentary opening created by counterbalancing strong personalities against diverging interests could not hold. By the time the initiative came to public attention, the arrangements that protected the program had already started to break down. When they collapsed, the fragile space could not be reopened. From its very beginning, the tensions present in the Lovelace testing program foreshadowed how it would end.

"Our First Girl in Space"

Lovelace and his counterparts began testing women for space fitness because various scientific studies suggested that women might make more efficient astronauts than men did. Women regularly outperformed men in enduring cramped spaces and withstanding prolonged isolation. In 1959 Colonel John Stapp, director of the Air Force Aeronautical Laboratory at Wright Field, complained publicly that men had not fared well in Wright Field's isolation tests. An unmarried woman performed quite well, however. A pair of international studies bore out his findings. In Great Britain, a woman withstood four days in an isolation chamber, besting any of the men tested at the same facility. In Canada, Wilma Sanson endured six days in isolation without suffering hallucinations—something none of the men tested there achieved. In 1961, University of Oklahoma Medical Center researchers found that women exhibited psychological coping patterns that predisposed them for success in sensory deprivation. Scientists suspected that women might be better suited to withstand a space capsule's confinement than men were.[1]

The most significant factor in women's favor, however, was weight. When loading a spacecraft, every added pound (whether in the capsule, astronaut, supplies, or hardware) required additional fuel to put it into orbit. In turn, that fuel itself added weight, necessitating even more propellant to lift the entire assembly off the pad. (During the Apollo moon launches, launching each pound of added payload required three additional pounds of propellant.) On the average, women are smaller and lighter than men and also consume less food, water, and oxygen than men do. The weight savings achieved by having a female astronaut would cascade through the entire design, lessening not only the weight of the pilot but also that of the supplies needed to sustain her and the fuel required to launch her.[2]

Female astronauts offered other advantages as well. Women suffer fewer heart attacks than men do. Scientists speculated that a woman's internal reproductive system would be more protected from radiation than men's external sexual organs were, and therefore less susceptible to damage from vibration or violent shaking. By 1962 a scientific study even asserted that a woman's skin was thicker than a man's. The speculation followed that female space travelers would be "less susceptible to space radiation [and] heat." If women could perform the same tasks in space as men could, the savings in weight made them attractive potential space travelers.[3]

The prospect of investigating these advantages—and challenging the taboo against subjecting women to undue risk—inspired *Look* magazine to tap into the curiosity with a publicity stunt. For four months, "a *Look* team" worked with NASA specialists assigned to Project Mercury, the space agency's human spaceflight program, to find out what scientists thought about women's potential for space travel. At the same time, aerobatics flying champion Betty Skelton put the theories to the test by undergoing physical testing and spaceflight simulations at the facilities where NASA trained its own hopeful space travelers. Skelton even worked out with the seven celebrated but as yet untested Mercury astronauts. *Look* featured the results in a February 1960 cover story under the provocative title, "Should a Girl Be First in Space?" The cover photo showed Skelton posed in front of a space capsule wearing an elaborate silver space suit, the visored helmet tucked under her arm. Inside, *Look* editors presented Skelton's foray into the life of an astronaut in a photo essay.[4]

The months-old American space agency gave *Look* unfettered access to its facilities and personnel. During a six-week period in October and November of 1959, Skelton traveled to five states to experience astronaut testing and training at NASA facilities and military sites. Skelton underwent physical testing at the Air Force School of Aviation Medicine in San Antonio, Texas. In St. Louis, Missouri, she visited McDonnell Aircraft Corporation to try on a space suit. During her time at Virginia's Langley Field, she examined the first American capsule to have been successfully recovered after being fired into space. There she also went through exercises with NASA's seven new astronauts. *Look*'s photographs showed her operating an orbital flight simulator while Alan Shepard and Wally Schirra stood by to offer "astronauts' guidance." She even donned an aqualung to swim with the astronauts in the indoor pool used to simulate weightlessness. According to one of the article's photo captions, Skelton got along so well with the seven pilots turned astronauts that she earned a nickname: "She was dubbed 'No. 7½' after she flipped the astronauts for morning coffee and beat them in the final turn of the coin." Whether the magazine's writers invented this moniker or the astronauts actually bestowed it on her, it did not do justice to her full-sized talents.[5]

The *Look* editors chose Skelton as the subject for the article because, except for her sex, she possessed the credentials to make her a competitive test subject. Her flying had earned her the nickname "the First Lady of Firsts." As an aerobatics pilot, Skelton became the first woman to accomplish an inverted ribbon cut. (In this maneuver the pilot cut a ribbon strung ten feet above the ground using the airplane's propeller—while flying upside down.) Skelton's other aviation records included piloting the smallest airplane to cross the Irish Sea, setting two light-plane altitude records, and winning three consecutive aerobatics championships. By 1951 she had retired from aerobatics at age twenty-five, although she continued to fly. Skelton also raced cars and jumped boats. In addition to nine sports car records, she set the women's world land speed record four times, becoming the first woman to officially drive a vehicle over three hundred miles an hour. Skelton still holds more combined aviation and automobile records than anyone else in history. In 1959 her résumé made her an attractive subject for *Look*'s experiment.[6]

As a potential astronaut candidate, Skelton combined lightning-quick reflexes with a disciplined personality. In her aerobatic flying, she accomplished her feats by practicing each maneuver for hours until she knew every move perfectly. Furthermore, she calculated safety guidelines for each stunt that included a 10 percent allowance for extra airspeed and altitude. She also exhibited resolute courage. When the NASA physicians asked Skelton to join the underwater weightlessness exercises, she got into the pool without saying anything, even though she could not swim.[7]

Strong and fearless, Skelton stood only five feet, three. Combining her physical talents with her compact size, Skelton offered the perfect reason why a woman should be first in space because she exemplified women's most obvious advantages: relatively small size and low weight. In 1959, when heavy lifting capacity challenged rocket scientists, women astronauts offered an attractive option. Not only would smaller launch vehicles be easier to engineer, they would also cost less. After putting Skelton through her paces for more than a month, *Look*'s writer concluded that she confirmed women's potential advantage for space travel: "She is just a petite example of the anatomical fact that women have more brains and stamina per pound than men." But if a compact woman was to go into space—first or otherwise—she would have to possess certain characteristics.[8]

Working with information gleaned from NASA's physicians, engineers, and psychologists, the article presented a composite vision of the first woman in space. In the researchers' best estimation, size remained the paramount consideration. As *Look* reported, "Our first girl in space will probably be a flat-chested lightweight under 35 years of age and married." Youth would ensure fitness and vitality, while marriage implied stability. In combination with these demographic requirements, the scientists also wanted the ideal female space traveler to possess a rather incongruous set of talents. Although the profile specified that "she will be a pilot," she would also be asked to leave any actual flying to her husband. "Her first chance in space may be as the scientist-wife of a pilot-engineer." In addition to holding a pilot's license, then, researchers wanted the female astronaut candidate to have scientific expertise in areas "rang[ing] from astronomy to zoology." That knowledge would allow her to assist the other members of the crew; scientists assumed that she would not be going into space alone.

Along with the unusual set of job skills, space researchers listed a complex psychological profile for the first woman in space. In addition to requiring that she maintain a stable marriage, experts wanted a woman who could act as a stabilizing force for all the other astronauts. According to *Look*'s composite, the first woman in space "will adjust well to isolation and be able to 'hibernate,' but also snap into immediate alertness. Her personality will both soothe and stimulate others on her space team." Fears that a woman's presence would disturb a mission's teamwork led scientists to require a higher standard of crew integration from the prospective female astronaut. In addition to being a team player who would not disrupt the mission, she also needed to take on the quasi-maternal responsibility of maintaining others' well-being.

Much of *Look*'s description focused on minimizing the problems that scientists saw in female physiology. In addition to starting the description by defining the prospective female astronaut as ideally "flat-chested," the profile stated the requirement a second time: "She will not be bosomy because of the problems of designing pressure suits." The ideal female candidate had to be able to fit the existing technology that had been designed for men. According to the physicians quoted in *Look*, women's reproductive cycles could not be easily accommodated in space either. Scientists planned not just to manage but to "eliminate" menstruation, using medication. Furthermore, the woman astronaut had to be "willing to risk sterility from possible radiation exposure." The scientists polled by *Look* assumed that men's biological needs posed challenges that had to be met but that women's bodies merely added unwelcome complications. Although a woman might potentially be included in a spaceflight, her physical femininity had to be minimized or even eliminated in order to make her suitable.[9]

Researchers' concerns about putting a woman into close quarters with men remained unspoken but were undeniably present in this description. Space scientists sought to assure themselves that the woman who would be locked inside a cramped spacecraft would be the boyishly built wife of one of the crew members, not a buxom single woman who might bring her sexuality on board with her. The close confines of a space capsule could not tolerate sexual tension. Furthermore, by defining the female astronaut as the soothing yet stimulating pilot-scientist-wife of another crewmember, NASA scientists signaled that they did not expect to find one easily. The unattainable job description listed in *Look* would not elicit qualified applicants anytime soon. Indeed, NASA allowed Skelton access to its testing facilities not because the agency wanted to find out how she performed but because it offered NASA a novel way to showcase its astronaut training for the public.

Although *Look* sponsored Skelton's unprecedented testing in order to sell magazines, the chance for a woman to prove herself on such tests remained significant, even if no one took the results seriously. After World War II, most women could not take advanced aerospace laboratory tests, let alone obtain access to the latest jet airplane or space simulator. Government air research centers hesitated to spend resources on female pilots who could not return the investment through flight service. Without an organized research program to foot the bill, the costs of access remained prohibitive. As a result, Skelton's experience marked one of the first instances when researchers examined a woman's capabilities using sophisticated aerospace diagnostic and simulation technologies.

Skelton's adventures showed just how uncommon—and outright unexpected—was a woman's presence at advanced aeromedical facilities. At Brooks Aerospace Medical School, since a female test subject had never been anticipated, the school could not provide anything for her to wear. With no clothes or footwear small enough for her, Skelton spent her visit wearing a tightly belted and rolled-up man's jumpsuit and her own high-heeled dress shoes. At another site, *Look*'s photographs show her having kicked off her shoes and stuffed her full skirt into a spinning testing chair. Either way, the lack of appropriate clothes visibly marked Skelton as out of place. Indeed, the *Look* photo spread's interest relied on showing a petite woman taking on oversized men's challenges.

Likewise, her singular experience stood out at the Naval Acceleration Laboratory in Johnsville, Pennsylvania. As the technicians prepared Skelton for her spin in the human centrifuge, they remarked that her ride would fit well into a popular Maidenform advertising campaign featuring women who dreamed about doing something highly unusual while wearing the maker's brassiere. Skelton recalled, "As they were putting me in the capsule, . . . one of them said, 'Golly, wouldn't this make a great ad? I dreamed I rode the centrifuge in my Maidenform bra!'" The good-natured joking highlighted how exceptional her presence was. Whether or not the *Look*-sponsored experiment had been intended to consider women seriously, the chance to be tested allowed Skelton to prove that women could tolerate these physical trials.[10]

Skelton recognized her unique situation. "I knew at the time they were not considering a woman really. . . . But [in] what little time there was associated with the NASA test and the astronauts, I did everything I could. I felt it was an opportunity to try to convince them that a woman could do this type of thing and could do it well." Getting a foot in the door meant a chance to prove that a woman could succeed and thus to influence scientific assumptions about women's capabilities. Within weeks of Skelton's specially arranged aerospace tests, another well-known woman pilot accepted an invitation to test her mettle in advanced aerospace simulations. This time the opportunity came courtesy of the U.S. Air Force, and this time the researchers had some real interest in the results.

Sometime in the fall of 1959, probably in October or November, Ruth Nichols, a fifty-eight-year-old aviation pioneer, traveled to Dayton, Ohio, to sample "some of the Astronaut tests" at the Wright Air Development Center. Given that *Look* magazine staged Skelton's experiments as fodder for its photo spreads, Nichols's experience recorded the first real scientific interest in testing a woman's physiology for space worthiness. Yet because researchers still could not conceive of a woman as a serious astronaut candidate, they shied away from discussing the subject publicly. If Wright Field's aerospace physicians had been interested in publicizing their investigations, however, they could not have found a better test subject. Nichols ranked as one of the most accomplished pilots in Earhart's and Cochran's generation of wealthy female flyers.[11]

During the early 1930s, in an era when flying enthralled the nation, Nichols became a darling of the aviation world. Born in 1901, she learned to fly at Wellesley College and had already become a recognized pilot by the time she set a 1929 record as only the third person—and the first woman—to land an airplane in all forty-eight states. In 1931 and 1932 alone, Nichols set the world altitude record, a transcontinental speed record, and the women's speed and distance records. The New York and New England Airlines tapped into her fame in 1932 by hiring her as the first female pilot for a passenger airline. That same year, her frustrated efforts to become the first woman to cross the Atlantic Ocean solo brought national attention to the "society-girl pilot." During one attempt, a spectacular crash cracked five of her vertebrae. The accident, combined with subsequent bad weather in the North Atlantic, kept her from achieving the record. Undaunted, she set another distance record before she had even recovered, clad alternately in a cast and an "iron corset" to support her still-healing spine.[12]

When air force researchers invited her to Wright Field in 1959, however, they were not reaching back into the past to resurrect a 1930s society-girl pilot; Nichols remained a prominent force in aviation throughout the 1940s and 1950s. In 1940, inspired by her Quaker roots, she founded Relief Wings, a charitable organization that used airplanes as emergency air ambulances. When World War II began, she shifted the organization's assets and her own humanitarian efforts to the Civil Air Patrol. In 1948 and 1949 she set a record for circumnavigating the world as a part of a crew, using the flight to raise awareness and funds for UNICEF (the United Nations International Children's Emergency Fund). In 1957 her autobiography, *Wings for Life*, familiarized the American public with her accomplishments again. Not content to reflect on a life of past achievements, however, Nichols continued to set aviation records. In 1958 she registered a women's altitude record of 51,000 feet in a supersonic air force jet. Even at fifty-six years old, "the dean of U.S. women fliers" could still perform at an elite level.[13]

A year after setting her jet-propelled altitude record, Nichols arrived in Dayton, Ohio, to undergo "astronaut tests" using state-of-the-art aeromedical equipment. She participated in weightlessness, isolation, and centrifuge experiments. Although she had previously broken her leg as well as her back in airplane crashes, her flying experience helped her to succeed on the tests. In a weightlessness simulator constructed out of a platform suspended on jets of steam, Nichols found that she could steady herself easily using a gyroscope despite the strain the fifty-pound mechanism initially caused. She likened her experience in the centrifuge to her first loop in an airplane. "Well, it's the same thing as what I said about being nauseated in an airplane: if you are busy doing something, your mind is completely taken up with that. Therefore, you don't think about the sensations that you are experiencing." Furthermore, she found that the presence of mind she demonstrated when her airplane ditched in the Irish Sea also helped her to endure the isolation experiments. Throughout the tests, she translated her aviation skills into success in the space simulations.[14]

Encouraged by her performance, Nichols prodded the air force researchers to pursue women's potential as astronauts more actively. "I put in a very strong urge that women be used in space flights. When I was out at Wright Field, they thought of this with horror, and they said under no circumstances." According to the responses she received, the aerospace physicians' reluctance had less to do with contemporary attitudes about women than with the lack of a physiological baseline. As Nichols recollected, "It had nothing to do with chivalry. The reasoning, according to the scientists with whom I spoke, was because they knew nothing about a woman, physiologically, which I thought was an extraordinary statement. . . . But I suggested a crash program to find out how a female reacted." The researchers expressed no interest in following up on her proposal. In the end, the premature announcement of Nichols's own successful examinations ended any chance that her test results would become the first component of a women's physiological baseline for space. Unknown to her, the announcement also served to halt another air force initiative before it got out of the planning stages.[15]

Project WISE

The primary advocate for that experiment was Brig. Gen. Donald Flickinger of the Air Research and Development Command (ARDC), the air force branch charged with research and development. In the years after World War II, he often collaborated with his old friend and colleague Randy Lovelace. During the late 1950s, Flickinger served as the air force liaison for the secret U-2 spy plane project, working with Lovelace and the Central Intelligence

Agency (CIA) to ensure the program's success. As they worked together to explore the highest reaches of the atmosphere, Flickinger demonstrated that he shared both Lovelace's research interests and his willingness to take chances in order to pursue them. Flickinger had made his reputation in 1943 by voluntarily parachuting into the dense Burmese jungle to rescue downed pilots trapped at an inaccessible crash site. Within the air force, he encouraged cooperation between the feuding military space programs. As a researcher and an administrator, "Flick" combined a passion for exploring human physical capabilities with a willingness to look past accepted boundaries. In 1959 his characteristic boldness led him to pursue investigations into women's space fitness even though he knew his superiors would not approve.[16]

Lovelace collaborated with Flickinger on this project because he had a vision of women as potential space travelers that—although ahead of its time in many ways—still replicated a gendered division of labor. Years before any human being flew in space—even a single pilot flying a suborbital trajectory that shot up into space and came right back to Earth—Lovelace envisioned orbiting laboratories populated by mixed-sex crews. On the one hand, his ability to foresee the potential for men and women flying together in space made him a part of a visionary group imagining the spacefaring dream: human beings living and working in space. On the other hand, the particulars of that futuristic dream remained bound by late 1950s conventions. As he envisioned astronauts working in space, Lovelace anticipated that women would be needed as laboratory technicians or communications officers, space-age equivalents of earthly laboratory assistants or telephone operators. Lovelace predicted women's participating actively in spaceflights, but his ideas still had them in pink-collar jobs.

As scientists who wondered if women could fill those positions, Flickinger and Lovelace shared the expertise, resources, and independence needed to begin a sustained research program. In his work with NASA's original astronaut selection in 1959, Lovelace had perfected a testing regimen for space fitness. Flickinger controlled some air force research funds. They just needed a suitable candidate to begin their investigations. They approached a successful young female pilot, Geraldine "Jerrie" Cobb, to be the initial test subject.

Their first meeting came in an unusual setting. The doctors met her while all three were in their bathing suits after an early morning swim at Miami Beach. They had come to Florida for the September 1959 Air Force Association meeting, and Tom Harris of Rockwell Aero Commander, whom Cobb had worked with to set a number of aviation records, made the introductions. Intrigued by meeting such an accomplished female pilot, the doctors invited Cobb to join them at the Fontainbleau Hotel to discuss recruiting women to take the Project Mercury tests. They asked if Cobb wanted to participate. She did. Lovelace and Flickinger needed to confirm Cobb's credentials, but they promised to contact her if her flight records proved adequate. When they checked her background, they found an impressive flying history.[17]

Cobb held commercial flight credentials and several flying records. Born on March 5, 1931, she began flying at age twelve when her father purchased a surplus Waco biplane. At sixteen, the earliest a pilot could legally solo, she earned her private pilot's license. She steadily added a flight instructor's rating and an instrument rating. Off to a fast start, Cobb became one of the youngest entrants in the All-Woman Transcontinental Air Races. By age nineteen, however, she found herself in a predicament familiar to female pilots. She had superb credentials, but no one would hire her. Finally she got a big break when Jack Ford, president of Fleetway, Inc., hired her to ferry aircraft to South America. For the next few years, Cobb flew all over the world, accumulating hundreds of hours in the sky. When that job ended, the Aero Commander Division of Rockwell-Standard supported her attempts at setting records, including an absolute altitude record of 37,010 feet set in an Aero Commander. By 1959 Cobb had earned recognition from the Women's National Aeronautical Association as the 1959 "Woman of the Year in Aviation" and the National Pilots Association as the "Pilot of the Year."[18]

The opportunity that Flickinger and Lovelace offered—to be tested for astronaut fitness—thrilled her. When the two aerospace researchers explained their plans, however, Cobb had no way of knowing how completely the resulting program would transform her life. In later speeches, Cobb recalled that the doctors had named the program Project WISE (Woman in Space Earliest). The initiative's inspiring title, sometimes also given as Woman in Space Soonest (WISS), derived from the name of the air force's rather less fortunately acronymed space project, Man in Space Soonest (MISS). When Flickinger went back to the air force to implement the plans, however, he simply called it a "girl astronaut program."[19]

The project aimed to test a group of female pilots. In preliminary discussions with Flickinger, Cobb suggested names of women she thought would qualify. Using the records of the Civilian Aeronautics Authority (CAA, the precursor to the modern Federal Aviation Administration) Flickinger researched aviation medical records, evaluating the women's qualifications based on age, height, weight, and flying experience. A group of eight potential candidates emerged. Other sources suggested the names of eight more women whose records, when checked at the CAA, might qualify them for participation. By the end of November 1959, however, before the second set of eight potential candidates had even had their flight records examined, all this work appeared to have been in vain. Flickinger's fellow ARDC researchers voted to cancel the program. Project WISE ended before it even began.[20]

Flickinger immediately wrote to his collaborators, Cobb and Lovelace, to explain why the initiative had been abandoned so suddenly. In a letter notifying Cobb about the setback, he commiserated with her: "Realize that I am even sorrier than you [are] on the unfavorable turn of events in my original plans." He expressed heartfelt disappointment but offered only a cryptic account of why the ARDC had withdrawn its support. Flickinger explained that "the unfortunate 'Nichols' release did much to 'turn the tide' against Air Force Medical Sponsorship of the program, and to this day I cannot find out the individual responsible for approving the release." Someone had made a public announcement about Ruth Nichols's testing.[21]

The public release revealing that Nichols had undergone astronaut examinations using air force equipment soured the atmosphere for further research on women. Nichols's presence at Wright Field implied that air force officials supported the idea of female astronauts, which they did not. The air force would not follow up on her examinations. Tragically, Nichols herself did not live to see any later program for testing women. Less than a year after she returned from Dayton, she took her own life in her New York apartment. Her suicide occurred nine months after the announcement of her examinations ended air force sponsorship of women's testing in December 1959—far too late for her untimely death to affect the ARDC decision. Some friends speculated, however, that disappointment about finding space closed contributed to her despondency.[22]

At the ARDC, the public announcement made air force researchers realize how unprepared they were for the political implications of testing women. In the face of the release, the ARDC summarily pulled its resources from Project WISE. For Flickinger, the mere mention of Nichols's testing meant that the female astronaut testing program faced considerable opposition. Those who opposed it offered several reasons, explanations that highlighted the pressures facing any women's space testing program in 1959.

Concerns about public reactions played a large role in his colleagues' decision. As Flickinger explained, "The concensus [*sic*] of opinion . . . was that there was too little to learn of value to Air Force Medical interests and too big a chance of adverse publicity to warrant continuation of the project." Aerospace researchers worried that examining women left them vulnerable to accusations of frivolous research. Having finally gotten the hard-won permission to find a real-life Buck Rogers, they hesitated to do anything to jeopardize that goal. Laboratory managers also felt considerable pressure not to use expensive equipment unless the results could be applied directly to human spaceflight (or to maintaining America's combat readiness). Without a sanctioned project intending to use female astronauts, the details of women's performance did not seem worth knowing.

No less significantly, the ARDC researchers' concerns about "adverse publicity" reflected postwar beliefs that women should be protected, not shot into space. Their thinking echoed other contemporary assessments of female astronaut investigations. *Look*'s Skelton article also suggested that "many [experts] believe that women's biggest obstacle to being first [in space] is our cultural bias against exposing them to hazardous situations." Air force officials did not want to risk the repercussions. After the Nichols release, ARDC researchers concluded that Project WISE could not continue without press attention—and any such publicity would likely be negative.[23]

In addition to concerns about publicity, the ARDC scientists also had material considerations in mind when they canceled Flickinger's Woman in Space Earliest project. They objected to the expense of modifying partial pressure suits to fit women's figures. Partial pressure suits allowed pilots to withstand the extraordinary forces exerted by high-performance flying. In complex aviation maneuvers and the tests that simulated them, increased gravity loads caused the subject's blood to rush into or out of appendages and, more dangerously, disturbed the blood flow to the brain. When blood rushed to the head, the pilot would "red out" as vision diminished and pressure made it harder to function. When blood rushed out of the head, the pilot would "gray out" as dizziness led to unconsciousness. Although many fliers learned to tense their muscles to keep blood flowing to the brain, that technique had obvious limitations. By exerting pressure on the pilot's body, a pressure suit offered a remedy to the fainting or disorientation caused by blood rushing throughout the body. A partial pressure suit, encasing the torso, upper thighs, and upper arms, permitted the subject to endure strenuous maneuvers.[24]

Given that the ARDC did not expect to learn much "of value" from testing women, the cost of refitting partial pressure suits to fit women proved too high to consider. Having women take the tests without the supporting garment would invalidate any results because they could not be compared with performances enhanced by partial pressure suits. As Flickinger wrote to Lovelace, "One of the major objections made by Tux [Turner] was that we could not justify the expense of altering the PPS's [partial pressure suits] to fit the girls and without the PPS, . . . there wouldn't be too much for correlation and I agree with him." Tux Turner and the other ARDC researchers apparently missed the irony that the David Clark Company, the firm that made the partial pressure suits, also manufactured brassieres. Expense may still have been an issue, but if the ARDC researchers had wanted to adapt the suits, those charged with altering them had a wealth of experience in fitting women's curves.[25]

More than anything else, the possibility of women astronauts in 1959 foundered on the obstacle created by the space research community's profound discomfort with women's physiology. Despite medical data collected from the armed services' women's divisions during World War II, aerospace scientists still thought they knew little about the particular workings of the female body. Moreover, the complete absence of a physiological baseline shamed no one. Like other medical researchers of the time who assumed that women's monthly cycles changed their bodies so fundamentally that they became unreliable test subjects—and yet prescribed remedies for women based on all-male studies without any adjustment—aerospace scientists simultaneously declared women to be both too complicated and largely irrelevant.

Because Flickinger concluded that he could not overrule the "great unanimity of opposition" expressed within the air force, he and Lovelace needed another way to pursue the project. After the air force pulled its support, Flickinger asked Lovelace to adopt Project WISE at his private Foundation. He agreed. Flickinger forwarded the CAA records of potential candidates compiled during the discussions with Cobb. At the same time, Cobb's own testing progressed under Lovelace's guidance.[26]

Despite all the delays that hindered Cobb's participation in the astronaut fitness tests, she remained optimistic about their possibilities. After her early December letter of apology from Flickinger, she heard from Lovelace by Christmas 1959 that her candidacy had been approved. She would be notified after the first of the year about reporting to Albuquerque. Finally, in February 1960, six months after the aerospace physicians first approached her, Cobb underwent the Lovelace Foundation's Project Mercury physical examinations.

She performed beyond expectations. Quietly, and without any public notice, Cobb became the first woman to pass the examinations that the Lovelace Foundation designed to screen astronaut candidates for NASA.

In hindsight, this quiet period must have seemed like the calm before the storm. Her achievement became international news on August 19, 1960, when Lovelace announced her results at the Space and Naval Medicine Congress in Stockholm, Sweden. Female pilot Jerrie Cobb had passed the same physical examinations used to qualify the Project Mercury astronauts. The Associated Press news wire picked up the story, spawning articles in newspapers throughout America. The public announcement of Cobb's success thrust her into the spotlight.[27]

Not only had Cobb been successful, her test results had demonstrated that women might be better suited to travel in space than men were. As Lovelace declared in Stockholm, "We are already in a position to say that certain qualities of the female space pilot are preferable to those of her male colleague." This did not mean, however, that women would become astronauts anytime soon. Below a photograph of Cobb smiling at the camera, the *New York Times* report of the announcement ended with a cautionary note. "Dr. Lovelace said in the lecture at the Space and Naval Medicine Congress that the first female space flight was still far off and that there was no definite space project for the women." Nonetheless, the press identified Cobb as the first successful female astronaut candidate.[28]

The next week's issue of *Life* magazine featured an exclusive photographic essay describing the testing that proved Cobb "Fit for Space Flight." The pictures depicted both the astronaut testing—Cobb lying on a tilt table having her blood pressure checked or breathing into an elaborate apparatus that measured her oxygen use—and Cobb's personal life: Cobb flying her airplane, swimming, praying, or playing tennis. *Life* described her as "the first prospective space pilot in a hitherto unannounced 12-woman testing program." Unlike *Life*'s straightforward reporting, however, much of the initial coverage of Cobb's achievement revealed the tensions surrounding women's issues in the early 1960s.[29]

As a woman who had proved herself as fit as the most revered American heroes, the Mercury astronauts, Cobb presented an obvious challenge to the idea that female physiology naturally limited women's potential for public roles. Recognizing a story that resonated with a public excited by exceptional women, the media featured Cobb's feat in newspapers and magazines. Without widespread feminist consciousness challenging stereotypical characterizations of women, however, the press struggled to report a woman's physical prowess in the language of the women's page. The resulting coverage heralded Cobb's achievements while simultaneously reducing her to an acceptable feminine role.

Accounts of Cobb's astronaut testing fixated on traits that had not been deemed newsworthy when equivalent stories had been filed about NASA's male astronauts. The *Washington Star* covered Cobb's success as if she had won a beauty contest, not passed astronaut tests, reporting in detail on her weight and proportions: "Miss Cobb, who has a 36-26-34 figure, said she lost 7 pounds during a week of testing but regained them. She stands 5 feet 7 inches and weighs 121 pounds." *Time* magazine blended her measurements with other traits and touted her aviation achievements but found it necessary to invent a feminized word for astronaut: "The first astronautrix (measurements: 36-27-34) eats hamburgers for breakfast, is an old hand at airplanes, with more air time—over 7,500 hours—than any of the male astronauts." Over the next three years, Cobb found herself being called everything from "astro-nette," "astronautrix," and "feminaut" to "space girl." The term *astronaut* apparently carried such masculine connotations that even a potential female candidate for space travel required coining a new label.[30]

Articles began and ended with descriptions characterizing the twenty-nine-year-old pilot as girlish. Reports emphasized her hairstyle (a ponytail) and her fears (grasshoppers). An Associated Press wire story even reduced Cobb to her hair color: "A blonde who may become this nation's first spacewoman predicted today that men and women will make space flights together." Reflecting the contemporary pressure on women to marry, reporters identified her as: "Miss Cobb, who said she had no matrimonial plans" and "Bachelor-girl Cobb."[31]

While the media struggled to describe a woman's physical strength in the language of feminine appearance, Lovelace worked out the administrative details of turning Project WISE into the Woman in Space Program. Throughout the fall and into the winter of 1960, in the midst of his other projects, Lovelace implemented the plans he had inherited from the ARDC. Using the CAA records that Flickinger and Cobb sorted, Lovelace began to turn the canceled plans into a working program.

Funding was a crucial issue. With only one exception, the women pilots who participated in Lovelace's Woman in Space Program stood a generation or more removed from their affluent flying foremothers. Airfare and a week's stay in a hotel cut deeply into budgets already trimmed to the bone to pay for flying time. The candidates would need financial assistance to be able to take the tests. In addition, the examinations themselves cost money. The Lovelace Foundation needed to pay for running the equipment while also accounting for the time spent by physicians, nurses, and technicians. In order to direct the project independently, Lovelace relied on private funds. Ultimately, that money came from Jackie Cochran and her husband, Floyd Odlum.

"Some May Get the Chance"

Although Cochran donated almost all the funds used to support Lovelace's Woman in Space Program and became closely identified with the project as it became better know, she joined the program in an ambiguous role. Cochran's interests were different from Lovelace's. She saw herself as the natural leader of the program, if not its chosen representative. Indeed, during the first months of the Albuquerque testing, Cochran entertained the possibility that she might also be a candidate. As a group, then, the initial actors in Lovelace's Woman in Space Program possessed differing visions of the program's shape and potential. As a result, when the project eventually sparked political opposition, they disagreed about how to proceed. The way Cochran came to participate in Lovelace's project affected how she acted with regard to the subject of women in space for the next seven years.

Jacqueline Cochran joined Lovelace's experiment in November 1960, three months after *Life* magazine announced Jerrie Cobb's successful examinations and almost a year after Flickinger transferred Project WISE to the Lovelace Foundation. Lovelace had inherited the outlines of Flickinger's ARDC project, but many decisions still remained. When Cochran received Lovelace's letter about the project, she concluded (probably correctly) that he wanted to draw on her previous experience as the head of a large-scale women's training program, the Women Airforce Service Pilots. Lovelace probably did not realize, however, that Cochran also assumed she was being asked to assume public leadership of the program. She immediately wrote back, volunteering her insights into what she called "the plans, still in formative stage, for the 'Woman in Space' program." Calling herself Lovelace's "special consultant," she offered several recommendations.

Specifically, she suggested that Lovelace ease the requirements for the testing so that more women could qualify. Cochran envisioned a large, long-term program. Based on her experience with the WASP, she suggested that Lovelace "liberalize" the entrance criteria, especially in the early stages, so that the successful group of women candidates would be comprehensive enough to be "conclusive." In particular, she recommended testing a few women older and younger than the declared age requirements. She also advised against banning married women. She reasoned that since it would be a long time before any of the candidates would actually fly into space, having prohibitions about marriage needlessly complicated the program because a woman who wanted to marry would have to resign. Although changing the requirements in these ways meant that Cochran herself qualified as a candidate, she did not address that possibility directly.[32]

Only one month later, however, taking the tests became Cochran's explicit goal. She decided to become a Woman in Space candidate, not just the program's consultant-financier. Two days after Christmas 1960, she wrote to Lovelace to arrange her schedule for taking the astronaut tests during a pending visit to Albuquerque. As she consulted her busy appointment calendar, she suggested, "That means I can take some of the tests you have in mind on Monday or possibly on Tuesday morning. But I imagine that except for some special ones you have my record from previous tests." She definitely planned to take the astronaut fitness tests, not just a routine physical examination. Referring to the bicycle ergometer test used to measure endurance, she specified, "I'll take that bicycle test Monday." For Cochran, participating in the astronaut project simply required coordinating the appointments. With regard to the broader program, however, she worried that other issues demanded closer attention.[33]

Cochran's scheduling letter also hinted at her growing discomfort with the perceived leadership of the female testing group. In her initial suggestions to Lovelace, she had stressed that "care should be taken to see that no one gets what might be considered priorities or publicity breaks." The persistent news coverage about Cobb's achievements disturbed her. A friend had just mailed Cochran a clipping featuring an article about Cobb's recent progress. In her letter to Lovelace, Cochran questioned her friend about Cobb's movement into more advanced testing. "What is the 'isolation chamber' that Jerry [*sic*] Cobb talks about? Has she advanced to Wright Field or was this one of the routines of the medical tests which took place at Los Alamos?" Cochran's questions telegraphed deep concern. She did not like being out of the loop. If she was going to fund the testing, she did not want to learn about major developments from outside sources. She wanted to be recognized—within the program and without—as an integral part of Lovelace's project.[34]

Cochran's participation in the Woman in Space Program hit a setback beyond her control, however, when she could not meet the physical requirements. Sarah Gorelick Ratley remembered having her own appointment with Dr. Lovelace interrupted so that he could talk to Cochran about her test results. As she sat in the office, "Dr. Lovelace looked in her patient folder and remarked that she would not be a candidate for the space program due to a heart problem that had developed." Cochran did not take the news well. Ratley recalled, "The door was closed as I walked down the hall but I was aware of loud voices in the background." Cochran could argue with Lovelace, but she could not dispute his conclusions.[35]

In addition to her being in her fifties—well over the project's age limit—persistent health problems dogged Cochran throughout her life. A botched appendicitis operation during her teen years had left her with abdominal adhesions that sometimes doubled her over in pain without warning. Because repeated interventions failed to remedy the problem, the attacks plagued her for the rest of her adult life. She also underwent sinus surgery and three eye operations. As Cochran's personal physician, Lovelace held records of her health problems dating back at least as far as a 1949 surgery. He knew her conditions. She could not be a candidate; her participation would have to be solely as benefactor and adviser.[36]

In the years that followed, Cochran denied ever having undergone the astronaut tests. When a reporter asked her about that possibility, she responded unequivocally, "Let me repeat that I was not one of the 20 who took these tests at Albuquerque." Rather than citing the age limit as the reason for eliminating herself from the candidate pool, however, Cochran reported that since she was consulting on the project, she considered it unfair for her to compete with the others. Even so, she had not entirely given up the dream of flying into space. As she speculated to the reporter, "Maybe early some morning they will suddenly need or want an experienced female to take the seat in a capsule. If so, I hope I am around close by. I could be treated as an expendable." Although she knew that Lovelace's program could not fulfill her dream, she still hoped that she might somehow become the first American woman in space. [37]

Once Cochran's chances of being a test subject ended, her role in the Lovelace testing may have been somewhat unclear, but her aspirations were not. She wanted to be recognized as the benefactor of the program. By the spring of 1961, she got the public appreciation she craved when a national magazine carried an article written by her, publicizing her leadership in the

Lovelace Woman in Space Program. The article did transform her role, but in a way she did not expect.

On April 30, 1961, when readers across the country opened their Sunday newspapers, they found a pair of petite brunette twins smiling at them from the cover of *Parade*, a magazine included as an insert. The photograph showed the Dietrich twins in matching flight suits, holding jet helmets, over the headline, "Jan and Marion Dietrich: First Astronaut Twins." The two California pilots—Marion, a commercial pilot and writer, and Jan, a corporate pilot with eight thousand hours in her logbook—had been flying since they were twelve years old. Inside, a photo spread showed the twins undergoing the Lovelace Foundation physical examinations while Cochran observed.[38]

The article published under Cochran's byline began with a disclaimer: she did not foresee women flying into space for six or seven more years. Without an emergency or shortage, she explained, the government could not afford to train women as jet pilots or test them as astronauts. Private funding financed the testing pictured in the article. In addition, she wrote, "So far, this medical research for women is unofficial and is just a gleam in the eyes of doctors interested in aero-space medicine." No official program existed. Nonetheless, the article's tantalizing subtitle read, "Famed Aviatrix Predicts Women Astronauts within Six Years."[39]

The rest of the article also undercut Cochran's initial cautionary tone. In fact, the section titled "Qualified Pilots" read like a job advertisement:

> Our women are civilian volunteers who:
> - are under 35 years of age, in good health and less than 5 feet 11 inches tall;
> - are qualified pilots with substantial experience in the air, measured by hours flown;
> - hold a Federal Aviation Agency instrument rating and medical certificate.

The subsequent paragraphs listed some of the examinations required for prospective astronaut candidates. Although Cochran specified that none of the women who volunteered would be obligated to participate in any future program if they underwent this testing, she affirmed that "all of them, so far, want to fly into space and some may get the chance." That last phrase must have leaped off the page for some readers. An even more enticing proposal ended the piece. Cochran concluded with a remarkable offer: "If you are a qualified and healthy woman pilot and you want to take the first step, which takes six days, write to me care of the Cochran-Odlum Ranch, Indio, California. You might become the first woman astronaut who really earns that name." In the overall balance of the article, the offer for women to take real astronaut tests far outweighed any caveats about a slow expected timetable for an official program.[40]

Letters from women pilots started arriving in Cochran's mail. Elizabeth "Betty" Fulbright White, a member of the last WASP class, 44-10 (which never flew because of disbandment), wrote in hopes of taking part even though she knew she was too old. "I realize that I could not meet the qualifications required for the program, however, I would like to know what the possibilities are for participating in the program in another phase." Two other letter writers had families including as many as three or four children. All hoped to take advantage of this extraordinary opportunity. Even sixteen months after the *Parade* article appeared, one flight instructor wrote to Cochran recommending a former student because he recognized how infrequently such opportunities for female pilots occurred. Cochran sorted through the responses generated by the *Parade* article looking for likely candidates to pass along to Lovelace.[41]

Nancy Lyman's enthusiasm made her note stand out. "I am 27 years of age, 110 lbs. wet, healthy as an ox, and raring to go," she wrote. The wife of a Navy test pilot and the mother of three children, Lyman had earned her private pilot's license, accumulating three hundred hours of flying time. She had postponed her commercial rating exam to compete in the All-Woman Transcontinental Air Races that summer, but she anticipated getting the rating in August. She had already arranged child care in case she got a chance to take the tests and wanted to know what else she had to do. "If you could give me more details on the requirements for this program, I will work twice as hard in order to meet these requirements." Cochran put her name on a list of women who might be interesting to Lovelace.[42]

Another woman whose name made the list stated forthrightly that she hoped to participate because she wanted to be a part of something that Cochran led. Myrtle Thompson Cagle, a newly married pilot from Georgia, had admired Cochran for years. When she saw the *Parade* article, Cagle sent her a letter. As Cagle explained it, she had wanted to work for Cochran during World War II, but "I missed your WASPS, because I was too young and too short. Perhaps I can participate in your space program." Like many who read the *Parade* article, she saw Cochran as the controlling decision maker in the women's astronaut testing program. Cochran put Cagle's name on the short list of potential test subjects. In all, from the letters inspired by the *Parade* article, Cochran culled eight candidates' information to send to Albuquerque.[43]

Because Cochran planned to attend the annual Paris Air Show, however, she asked Odlum to forward the names to Lovelace. As requested, Odlum mailed the candidate list to Albuquerque with an explanatory cover letter conveying Cochran's rationale, "She asked me to add: (a) that three or four of these women seem fully qualified to take the tests (b) that the others may lack in certain requirements, as for example instrument rating, but they look like good material." Not all of the women listed met the Woman in Space Program requirements.[44]

In truth, Lovelace's guidelines proved difficult to meet. Younger women had a hard time accumulating the necessary flight time (one thousand hours) and advanced ratings. Older women with abundant ratings and sufficient logbook hours did not always possess the requisite physical fitness or personal flexibility (freedom from family or work responsibilities) needed to take the tests. Lovelace rejected Fay L. Davis, a pilot with qualifications similar to Nancy Lyman's, because she had fewer than five hundred hours of flight time. Lyman did not qualify for the same reason. Regardless of whether they received invitations to Albuquerque, however, as a group the women whose names Cochran forwarded shared one trait. They had written to Cochran, as the public leader of Lovelace's Woman in Space Program, hoping she could give them the chance to take astronaut tests, as she had offered hundreds of other female pilots the opportunity to fly military aircraft during World War II.[45]

"Jackie Is Rather Unhappy"

Lovelace probably did not think much about the precise authorship of the cover letter accompanying the eight candidate suggestions. However, Cochran's handing this task off to her husband hinted at her growing dissatisfaction with the responsibilities Lovelace offered her. In Albuquerque, the Woman in Space Program proceeded apace. Staff members at the Lovelace Foundation extended invitations, scheduled testing, and conducted examinations. In the meantime, however, Cochran's participation had been reduced to clerical work: screening and forwarding applications. That role did not satisfy her.

The conflict between Cochran's desire for recognized leadership and her ambiguous status in the Lovelace testing program came to a head at the end of May 1961. Even though Cochran and Lovelace had been friends for years, she did not approach him directly. Cochran knew when to use personal and political allies to intercede for her. After all, her husband served as president of the Lovelace Foundation's board of directors. Odlum wrote to Lovelace on her behalf.

In a remarkable May 31, 1961, letter marked "*Personal and Confidential*," Odlum forcefully voiced his concerns about Cochran's role in the women's testing project. One of Lovelace's successful female candidates had written to Cochran mentioning additional testing planned for June. "Once again this leaves Jackie up in the air," Odlum wrote. As in December, when she had learned about Cobb's isolation tests from an outside source rather than directly from Lovelace, the news of further examinations once again made Cochran feel she was out of the loop. The incident reawakened discomfort that had just begun to be put to rest. As Odlum described it, "Jackie is constantly asked about this women's program and is so much on the edges of it that she has to reply in rather vague generalities. She has the unhappy feeling that you don't really want her to be a part of the program." Cochran felt sufficiently unappreciated that Odlum had decided to step in.[46]

He recited a litany of slights and oversights. According to Odlum, Lovelace had started the program without consulting Cochran. The original publicity had not been brought to her attention. Lovelace had selected the original candidate group without her input. Even when Cochran observed the Dietrich twins' testing, Odlum related, she felt she had been kept in a "'detached' position." Finally, Lovelace had not done enough to connect Cochran with NASA or to include her in further plans for the women's program.

If the cumulative effect of this list gave Lovelace an indication of the breadth of Cochran's resentment, the next line drove home its depth: "In view of the close friendship between the two of you and her well established place in aviation," Odlum reported, "Jackie can't believe it was pure oversight on your part." He also laid out Cochran's speculations about why she was being shut out, including her suspicions that "a certain Air Force General who is prominent in air and space medicine and who for certain reasons may be wanting to push Jackie out of the picture" might be behind Lovelace's lack of action. Whatever the reasons, the powerful couple had had enough.[47]

Odlum dictated the terms of participation that would satisfy Cochran. "Jackie does not want to be around if you don't want her. If you do want her she should be kept advised of the statements and letters that are being sent to the candidates and consulted about how the group will be organized for the next step." She should also be consulted about the group's size, composition, and future. Essentially, Lovelace needed to keep Cochran fully involved in all aspects of the project from that point onward. Odlum ended the letter with a frank challenge to his friend: "If you have personal or other problems dealing with the point of this personal letter you should lay them on the table because, as I said above, Jackie is rather unhappy." Under his signature, Odlum listed a carbon copy to Cochran. She would return from Paris expecting changes. Lovelace had one week to decide how to pacify his most important supporters.[48]

Notably, Odlum did not mention money in the letter. He did not have to. He and Cochran contributed generously to both the Lovelace Clinic and the Lovelace Foundation for Medical Education and Research. The Woman in Space Program itself illustrated just how generously they gave. Cochran initially donated $500 so that women who could not afford the transportation and lodging costs could still participate in the "'Astronaut' Program for Women." In addition to that first grant, however, Cochran and Odlum also financed the Foundation's expenses, the remaining costs, and much more.[49]

Cochran and Odlum donated the funds for Lovelace's private women's testing project in an especially generous form. Rather than reimbursing specific costs, they made broad grants of stock. In February 1961, the Cochran-Odlum Foundation sent Lovelace 2,500 shares of Federal Resources Corporation stock, worth $7,900. They directed the Foundation to sell enough of the stock to recoup its costs for each candidate checked. Odlum and Cochran donated another 3,200 shares of the same stock in November, adding $10,601.33 to the Lovelace Foundation's coffers. Cochran and Odlum expected that the donations would pay the Woman in Space testing expenses with money to spare. Rather than asking that the rest be returned, they added instructions that it be retained by the Foundation to be used as Odlum advised.[50]

Cochran and Odlum's support of Lovelace's Woman in Space Program amounted to a significant percentage of the Foundation's grant income for research in aerospace medicine and astronautics. Overall, the couple contributed almost $19,000. In comparison, the Foundation's entire 1963–64 Aerospace Medicine and Bioastronautics grant and contract income totaled only a little over $55,000. Since the general figures surviving from the Lovelace Foundation indicate that its income increased each year in the early 1960s, such a comparison probably underestimates the Odlum-Cochran impact on this budget area. In addition to financial contributions, Odlum also offered his business acumen and public clout as the Lovelace Clinic's chairman of the board. Lovelace could not afford to lose such valuable benefactors.[51]

To complicate matters further, the relationship between the two families was more than just a collegial professional friendship; Odlum and Cochran remained dear friends of the Lovelace family. When the family vacationed, the Lovelaces often stayed at the Cochran-Odlum ranch in Indio, California. Through the years, Mary Lovelace and Cochran wrote long, intimate letters to each other. By 1949 the families had grown so close that the Lovelaces named their third daughter after Cochran and asked the world-famous pilot to be young Jacqueline's godmother.[52]

Cochran and Lovelace traded professional favors as well. To a couple with persistent health problems, Lovelace offered special treatment and attentive medical care. As early as 1949, Cochran made special trips to Albuquerque for surgery. When she fell ill while traveling in Spain during 1955, she so trusted Lovelace as her personal physician that she wired him her symptoms from Madrid. Lovelace also went out of his way for Odlum, who suffered from almost crippling arthritis throughout his life. In 1957 Lovelace flew to England to tend the magnate during a medical emergency. Throughout much of Cochran's life, either Randy Lovelace or the Lovelace Clinic took care of her health.[53]

In return, Cochran referred a number of wealthy patients to the Lovelace Clinic. She did so to help Lovelace's practice but also because she thought he could help friends she feared might otherwise end up seeing "a bunch of charlatans who like to make money off of rich people." From their first interactions regarding the Collier Trophy in 1940, Cochran and Lovelace's personal and professional friendship benefited them both. Now Cochran's dissatisfaction with her role in the Woman in Space Program jeopardized their relationship. Lovelace needed to make amends quickly.[54]

Lovelace's three-page response to Odlum's letter started slowly but ended with answers to each of Cochran and Odlum's complaints. Lovelace first justified his actions by explaining that some of the slights had been unintended: "At the time we initiated this program Jackie was in Europe so this explains why we did not tell her about it then." As the letter continued, however, Lovelace addressed each of Odlum's concerns more directly until the last page answered the objections point by point. Lovelace committed himself to involving Cochran throughout the program. "We will continue to send you copies of all the correspondence in connection with this program. When we have the names of the girls that will be coming for tests, we would be happy to have Jackie write them about smoking and exercise." To make good on this pledge, he enclosed a list of eight names and addresses of women who had already passed the examinations and of five women who were scheduled to be tested. Finally, Lovelace tried to mend personal fences with a handwritten postscript: "P.S.: I consider Jackie and you the couple I am closest too [*sic*] of all the couples I know." He not only wanted to save the project, he wanted to maintain their friendship.[55]

As it had in the past, Cochran's personal relationships and Odlum's financial clout solidified her influence. Although many powerful men used private networks to advance their careers, Cochran distinguished herself as a woman who excelled at using both personal and professional connections to promote her interests. Her own extraordinary favors cemented the attachments. Cochran's and Lovelace's initial meeting over the 1939 Collier Trophy exemplified the pattern. Throughout her life, Cochran had similar associations with aircraft manufacturers, military generals, and other influential men. No fewer than two U.S. presidents owed significant turns in their political fortunes to Cochran.

In 1951 Cochran helped draft the president. That year, the Republican Party floated Gen. Dwight Eisenhower's name as a desirable presidential candidate, but Eisenhower would not commit to running. To persuade him, his supporters—including Cochran—staged a "draft Eisenhower" rally at Madison Square Garden. They filmed the speakers, the "I Like Ike" posters, and most of all the overflow of people filling the arena on his behalf. The next day, Cochran herself flew the movie to Eisenhower in Paris. After a half-hour meeting with her, including a viewing of the film, Eisenhower's attitude changed. He began to plan for a return to the United States and a campaign for the presidency. For the rest of his life, Eisenhower maintained his friendship with Cochran, even writing his memoirs while living in a guesthouse on her ranch. Her 1951 actions cemented her first bond with a president of the United States. Four years earlier, however, Cochran had helped another aspiring national politician, who later became her second friend to attain the presidency. Long before she helped to draft Eisenhower, she permitted another man, a struggling Senate candidate, to remain in politics.[56]

In 1948 Cochran saved Lyndon Johnson's political life. When Johnson developed a kidney stone during his 1948 Senate primary race, he continued to make public appearances until a cripplingly painful infection and fever finally hospitalized him. Facing surgery and a six-week convalescence that would keep him from campaigning until just before the primary, Johnson nonetheless refused surgery and refused to leave Texas. Warren Woodward, one of Johnson's campaign managers, recalled Cochran's determined assistance: "I was taking calls at the hospital by this time, and she [Cochran] came on the line and in a very forceful, direct way she said, 'This is Jacqueline Cochran. I'm here in Dallas. . . . You just tell Lyndon that I'm going to take him [to the Mayo Clinic] in my [Lockheed] Electra.' . . . She didn't give me a chance to say yea or nay, just said, 'Give him that message.'" Over LBJ's objections, Woodward and Lady Bird Johnson checked the congressman out of the hospital. At Mayo, Johnson underwent a nonsurgical procedure that allowed him to return to the Texas campaign trail only two weeks later—with a full month left before the vote. Johnson's gratitude for Cochran's actions cemented her lifelong friendship with the future president. Fourteen years later, when the question of women astronauts generated national attention, Cochran called on that relationship.[57]

Cochran's bonds with powerful people often allowed her to do what she wanted. In the case of Lovelace's Woman in Space Program, Odlum's pressure on Lovelace reaffirmed Cochran's place in the program. After she returned from Paris, Cochran resumed working on the project on the terms Odlum had negotiated. Nonetheless, her initially ambivalent connection with the project affected how she acted. She never regained the enthusiasm she had demonstrated at first, when she thought she might participate as a candidate or when she hoped to lead the women's astronaut testing program as she had led the WASP. Frustrated by her diminished role, Cochran revised her memory of what happened to shed a more favorable light on her participation. Indeed, she literally rewrote the history.

In an "as told to" autobiography collaboratively written by Cochran and Maryann Bucknum Brinley, the authors assert that Cochran thought up the idea of testing women as potential astronauts and then collaborated with Lovelace to start the testing. In the book, Cochran recalled that she and Lovelace combed the records of the Ninety-Nines to compile a list of suitable candidates. Cochran's longtime personal secretary, Maggie Miller, echoed the memory, "Just for the heck of it, Jackie put herself into the same tests—medical and emotional—that they gave the astronauts back then in the '60s. Then, she had a whole program set up with Randy Lovelace, designed to prove just how capable women were." Miller's recollection says more about her loyalty to Cochran than about the research project's actual history. Over the years,

Cochran came to remember that she had initiated the Lovelace women's testing because this new memory fit better with the part she had wanted to play.[58]

The actual history of how Lovelace's Woman in Space Program began—as an air force project beset by concerns about adverse publicity, which later drew funding from a benefactor who became increasingly dissatisfied with her role—explains much more clearly why the program disintegrated when outside pressures increased. Flickinger and Lovelace created a space for women's testing to exist because they did not allow administrative structures to limit the scope of what they wanted to investigate. Intrigued by the possibilities of women's filling support roles on space missions and undaunted by bureaucratic opposition, they pushed ahead even after the air force pulled its support.

The women pilots who answered Lovelace's invitations never knew about the tensions that permeated the project from its inception. They did not realize that the Woman in Space Program had already died once, only to be resurrected under private sponsorship. They just knew that chances like this did not come along very often. Eager to put themselves to the test, they cleared their schedules when they received the coveted invitations to take the Project Mercury tests at the Lovelace Foundation. They could not have known what a physical and mental ordeal awaited them.

President Franklin Delano Roosevelt presenting the 1939 Collier Trophy. In 1940 Jacqueline Cochran (pictured behind the trophy) advanced her own career and cemented a lifelong friendship with Randy Lovelace when she lobbied for Dr. Walter M. Boothby, Dr. William Randolph Lovelace II, and Capt. Harry Armstrong (left to right behind Roosevelt) to be named, along with the airlines, as recipients of the trophy. *Source:* Lovelace Sandia Health System.

With the tank of the BLB oxygen mask strapped to his leg, Dr. William Randolph "Randy" Lovelace II prepares for a record-setting 1943 parachute jump to test the invention. Lovelace's high-altitude research would lead him to aerospace research—and women's astronaut testing. *Source:* National Air and Space Museum, Smithsonian Institution (SI 77-10380).

Astutely aware of how to use her public image to promote both her flying and her cosmetics business, Jacqueline Cochran applies lipstick while sitting in the cockpit of a P-35 Seversky Fighter before takeoff in the 1938 Bendix race. *Source:* Eisenhower Presidential Library.

Ruth Nichols, a "society-girl pilot" who became "the dean of U.S. women flyers," took astronaut tests at Wright Field in 1959. The premature announcement of her air force testing precipitated the end of Flickinger and Lovelace's Project WISE (Woman in Space Earliest). *Source:* National Air and Space Museum, Smithsonian Institution (SI 77-4169).

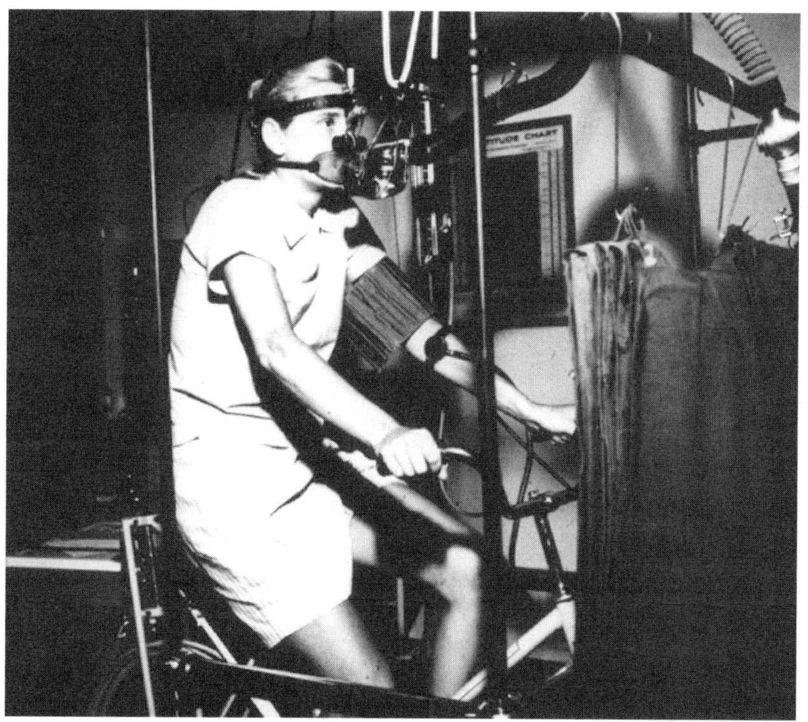

Shown here taking a stationary bicycle test that measured endurance and oxygen use, Jerrie Cobb became the first woman to pass the astronaut physical that the Lovelace Foundation developed for NASA. With only the addition of a gynecological exam, the women underwent exactly the same testing regimen that NASA's male candidates did. *Source:* Lovelace Sandia Health System.

Jerrie Cobb poses with a full-scale model of NASA's Mercury capsule on May 27, 1961. Later that evening, NASA administrator James E. Webb announced Cobb's appointment as a special NASA consultant. Two days earlier, however, President John F. Kennedy had committed the United States to a moon landing, a directive that streamlined the American space effort, making decision makers less likely to consider a female astronaut. *Source:* NASA photo.

THE VICE PRESIDENT
WASHINGTON
March 15, 1962

Dear Jim:

I have conferred with Mrs. Philip Hart and Miss Jerrie Cobb concerning their effort to get women utilized as astronauts. I'm sure you agree that sex should not be a reason for disqualifying a candidate for orbital flight.

Could you advise me whether NASA has disqualified anyone because of being a woman?

As I understand it, two principal requirements for orbital flight at this stage are: 1) that the individual be experienced at high speed military test flying; and 2) that the individual have an engineering background enabling him to take over controls in the event it became necessary.

Would you advise me whether there are any women who meet these qualifications?

If not, could you estimate for me the time when orbital flight will have become sufficiently safe that these two requirements are no longer necessary and a larger number of individuals may qualify?

I know we both are grateful for the desire to serve on the part of these women, and look forward to the time when they can.

Sincerely,

Lets stop this now!

Lyndon B. Johnson

File

Mr. James E. Webb
Administrator
National Aeronautics and Space Administration
Washington, D. C.

Vice President Lyndon B. Johnson's emphatic "Let's stop this now!" recorded his private opposition to women in space, an idea he feared would detract from NASA's missions and undermine its political support. Johnson's exclamation remained buried in his files for almost forty years. *Source:* VP Papers, Lyndon Baines Johnson Presidential Library, box 183.

On July 17, 1962, Jerrie Cobb and Jane B. Hart testified before a special subcommittee of the House Committee on Science and Astronautics. This widely published photograph showing Cobb's shoes off under the table undercut the force of her arguments by featuring her slip in feminine etiquette. *Source:* Library of Congress.

On the second day of the subcommittee hearings, astronauts M. Scott Carpenter and John Glenn testified that women could not be astronauts because they were not jet test pilots. The presence of Carpenter and Glenn, America's two most recent space travelers and new national heroes, reinforced NASA's arguments that the agency's success depended on being allowed to pursue space achievements without undue interference. *Source:* Library of Congress.

Seven of the thirteen women who passed the Lovelace Foundation's astronaut tests in the early 1960s gathered to watch NASA astronaut Eileen Collins's 1995 flight as the first female pilot astronaut. From left to right: Gene Nora Stumbough Jessen, Wally Funk, Jerrie Cobb, Jerri Sloan Truhill, Sarah Gorelick Ratley, Myrtle "Kay" Cagle, and Bernice "B" Steadman. *Source:* NASA photo (KSC 95PC-277).

5

Lovelace's Woman in Space Program

When *Life* magazine's October 24, 1960, issue arrived in homes and on newsstands, its readers received an update on Jerrie Cobb's continuing progress through space-readiness testing. Under the title "Damp Prelude to Space," a photograph taken with special infrared film showed Cobb floating calmly in a pitch-black pool, testing her tolerance for sensory deprivation. The article described this test of psychological mettle as "one of the most stern ordeals which an Astronaut must be able to withstand." Most of the forty volunteers whom Dr. Jay T. Shurley tested in the soundproof, light-proof tank in Oklahoma City lost touch with reality, at least temporarily. In contrast, the psychiatrist rated Cobb's endurance as "extraordinary." In nine hours and forty minutes, *Life* informed its readers, she demonstrated the adaptive abilities required to be "a space traveler who may have to react in split seconds after dreary hours of doing nothing." As she had at the Lovelace Clinic, Cobb passed the astronaut fitness tests with flying colors.[1]

In Fort Sill, Oklahoma, twenty-year-old Wally Funk, an accomplished pilot and a competitive skier, pored over the magazine's coverage of Cobb's accomplishment. In it she saw a chance to advance her own dream of going into space. "So, [I was] thinking, 'Oh! This is really what I want to do!' I mean, there weren't a lot of things back in the sixties for girls to do." After representing the southwestern United States in national slalom and downhill skiing competitions, Funk had become Oklahoma State University's Flying Aggie Top Female Pilot in both 1959 and 1960. An ambitious young woman who sought out opportunities to use her talents, Funk extracted Dr. Shurley's name from the article and wrote him a letter volunteering for his astronaut testing program.[2]

Shurley's response revealed the underpinnings of Cobb's testing arrangements. The formal program that Funk wanted to join did not exist. Unable to provide the access she requested, Shurley explained, "As of this moment, I have no official connection with any organized program for women Astronaut candidates. My work with Miss Cobb was on a purely informal basis." He suggested that she contact either Cobb or Lovelace, who had arranged the isolation trials, and he wished her luck. Buoyed by his words and armed with Randy Lovelace's address (which Shurley enclosed), Funk sent a letter to Albuquerque, offering to subject herself to the Project Mercury physical.[3]

When Lovelace inherited Project WISE from Donald Flickinger and the air force's Air Research and Development Command early in 1960, informal arrangements provided the best way to turn the proposal into a working program. As a result, throughout 1960 Lovelace negotiated the logistics of Cobb's testing ad hoc. She underwent sensory deprivation tests in Oklahoma City followed by advanced aeromedical testing at the Naval School of Aviation Medicine in Pensacola, Florida, thanks to permissions Lovelace set up individually. The "purely informal" arrangements allowed Lovelace to bypass the usual barriers that blocked women from aerospace experimentation. In addition, Lovelace avoided alerting those who had opposed testing women only eighteen months earlier.

Within his own Foundation, however, Lovelace had the means to begin a sustained investigation of women's potential for space flight. Despite squeezing the women's examinations into gaps in the Foundation's schedule, the Woman in Space Program projected the appearance of a project with potential. Just how much potential—and how far it might carry the participants—remained open to interpretation.

During the spring and summer of 1961, Lovelace's invitation presented an enticing challenge. One by one, each of the women pilots arrived in Albuquerque unsure of what to expect but ready to work hard to succeed. As they began the examinations, they wondered what some of the unusual tests measured and when they would ever have time to digest a real meal. In the end, the exacting regimen tested not only their physical capabilities but also their emotional endurance.

Examinations of Potential Women Astronauts

During the fall of 1960, Cobb talked to some of her fellow pilots, sounding them out to see if they had any interest in taking the Lovelace examinations. Jerri Sloan Truhill recalled that Cobb called her in 1960 asking whether she would be able to get away for a secret government program. Since she already worked on testing contracts that operated on a need-to-know basis, she did not ask for more information. When she received a letter from Dr. Lovelace, she recalled, "I thought, 'Well, this must be what Jerrie was talking about.'" When Fran Bera picked up a new twin-engine airplane from Cobb's Oklahoma employer, Cobb took the opportunity to approach her about the project. She described Lovelace's plans, probing whether Bera was interested in participating. She was. Soon after, Bera received a letter of inquiry from Dr. Lovelace himself.[4]

The invitations that Sloan and Bera received presented the first phase of testing as an organized program with impressive credentials. Sarah Gorelick (Ratley) received an identical letter about the project in late May 1961. It began, "We have been informed that you may be interested in volunteering for the initial examinations for female astronaut candidates." The offer carried the official letterhead of the Lovelace Foundation for Medical Education and Research and Randy Lovelace's signature. The Lovelace name alone lent prestige to the program.[5]

Anyone who followed the American human spaceflight program knew that "Lovelace" signified the best in human spaceflight research. When NASA announced the original Mercury Seven astronauts at a press conference on April 9, 1959, Dr. Lovelace sat on the dais alongside the astronauts as one of only six NASA representatives presenting the new space travelers to the public. When *Life* magazine profiled the tests used to select those men, Lovelace's byline appeared on the article describing the examinations conducted at the Lovelace Clinic. In the new world of human space exploration, the name Lovelace epitomized space medicine. Sending the invitation on Lovelace Foundation letterhead underscored the initiative's credentials.[6]

In the correspondence asking women pilots to participate in this testing program, Lovelace did not promise a women's version of NASA's Project Mercury. Rather, he emphasized the provisional status of the first stage of examinations. As the note to Gorelick stated, "These examination procedures take approximately one week and are done on a purely voluntary basis. They do not commit you to any further part in the Woman in Space Program unless you so desire." Although the request spelled out that volunteering for the tests did not lock the participant in to a long-term program, the note commented less clearly on what future the initiative itself might offer. Lovelace's cautionary remark about commitment to "any further part" in the program implied that subsequent phases might follow the initial examinations.[7]

Despite the presentation of the testing project as a credentialed program sponsored by the aerospace research clinic most closely affiliated with the official United States space agency—or the implication that the project might have future stages that candidates could participate in—Lovelace's inquiry did not promise a full-fledged women's astronaut program. None of the letters mentioned NASA or official United States space efforts in any way. Those who received invitations would not have thought that the women's testing was a NASA program. At the same time, however, the pilots Lovelace contacted certainly appreciated the significance of receiving a solicitation from the space agency's own de facto medical center.

In order to participate, each woman had to apply. Enclosed with the inquiry letter, Lovelace included a small notecard outlining "the qualifications of the women astronauts." To be evaluated for the program, each potential candidate needed to send the Lovelace Clinic a complete account of the information requested. The letter reassured each potential candidate that she would not be charged for the examinations. "Your only expenses would be traveling expenses and room and board at a nearby motel." Lovelace offered not only access to sophisticated aerospace testing but also the chance to take the most extensive physical evaluations available, solely for the cost of traveling to the clinic.[8]

Throughout the early months of 1961 and well into the summer, Dr. Lovelace sent inquiry letters to twenty-five women inviting them to undergo "the initial examinations for female astronaut candidates." Since letters went out on a staggered schedule, the process of offering initial invitations and receiving the responses stretched throughout the spring and summer.[9]

Not every woman who received an invitation accepted the offer. Marilyn Link, an accomplished corporate charter pilot and flight instructor whose brother Edwin invented the Link Trainer (the first aviation simulator), chose to decline. In her late thirties, Link felt she was too old to begin such a project. In addition, the invitation arrived while she was making a career change. Because she had not sought the opportunity, she decided not to accept it.[10]

The complete list of potential candidates included:

Passed Lovelace Tests	*Completed Lovelace Tests*
Myrtle "K" Cagle	Frances Bera
Jerrie Cobb	Virginia Holmes
Jan Dietrich	Patricia K. Jetton
Marion Dietrich	Georgiana T. McConnell
Wally Funk	Joan Ann Meriam [Smith]
Sarah Gorelick [Ratley]	Betty J. Miller
Jane B. Hart	*Declined or Unknown*
Jean Hixson	Dorothy Anderson
Rhea Hurrle [Woltman]	Marjorie Dufton
Irene Leverton	Elaine Harrison
Bernice "B" Steadman	Sylvia Roth
Gene Nora Stumbough [Jessen]	Marilyn Link (declined)
Jerri Sloan [Truhill]	Frances Miller

Five other names appeared on the schedule of those invited to take the tests but not on the list of women who completed the examinations. According to a roster of "Girls Tested" sent to Jacqueline Cochran in mid-June 1961, three women—Dorothy Anderson, Marjorie Dufton, and Elaine Harrison—answered Lovelace's letters agreeing to take the examinations. A handwritten notation indicated that Dorothy Anderson's testing would begin on June 23. She eventually canceled because she could not take the time off from her job as the sole full-time flight instructor at a large flight school. Perhaps a job or family conflict prevented the other two women from attending. Or possibly they traveled to Albuquerque only to have the Lovelace medical team discover a problem that ended their participation. A fourth woman, Sylvia Roth, who was scheduled to take the tests with her former flight instructor Irene Leverton, canceled because she could not leave her work as a corporate pilot for *Encyclopaedia Britannica.* The final potential candidate, Frances Miller, chose not to answer Lovelace's invitation. She had heard through the grapevine that the experiment had no future. Of the twenty-five women contacted by the Lovelace Foundation in early 1961, nineteen (including Cobb) completed the full set of examinations.[11]

The women who accepted Lovelace's invitation mailed their information to Albuquerque and got responses like the one that Jerri Sloan received on

March 24, 1961. Dr. Lovelace wrote, "We have reviewed the credentials you have sent in and find that you are acceptable for these examinations." Lovelace's form letter made no outright promises of a developed program. On the other hand, it contained several suggestive instructions.[12]

First, Lovelace called the group "potential women astronauts," not testing candidates or experimental subjects. Whether that "potential" depended on the candidate's own abilities or on larger questions of space policy remained unexplained. Second, without saying anything directly about the future of the program, Lovelace cautioned that the individual participants would have to observe group rules regarding press coverage. The note specified, "There will be no announcements or releases on this program until all of you have had your check-ups and then, with each participant's consent, only the names of those that pass the entire examination will be released." Such discussions of group press coverage suggested future attention from the media. Dr. Lovelace reinforced that impression when he expressed in his closing line that he "hoped to have the candidates that pass the examinations meet together late this spring." Successful participants could expect their test results to be newsworthy.

In addition to warnings about press coverage, Lovelace's response also outlined the financial support for the week at the Lovelace Clinic. Lovelace explained that Jacqueline Cochran had volunteered to assume the costs of any "expenses for room, food and transportation up to $200 during the approximately six days you would be here." Cochran's financial support came as a part of her involvement as the program's "special consultant." The women pilots who received Lovelace's offer could undergo expensive aerospace examinations at no cost to themselves beyond the time required to take the tests. At a time when prohibitive cost and exclusive access systematically excluded women from cutting-edge aerospace technology, having both barriers lowered at once was extraordinary.[13]

Finally, Lovelace's letter set the schedule for Sloan's testing. Lovelace requested that she arrive in Albuquerque the next Sunday to be ready for examinations that would begin early Monday morning. She needed to be there in less than one week. As she hastily made her travel plans, Sloan jotted notes about her arrangements right on her invitation, above the Lovelace Clinic's letterhead. As a mother of two and an active pilot, she had quite a bit to juggle to find time to participate in the testing. Even with the short notice, she made the adjustments necessary to try her hand at Lovelace's "examination of potential women astronauts."

As the Woman in Space Program continued into the summer of 1961, however, the notice each woman received became even shorter. Phone calls replaced letters. To take advantage of available openings in the Clinic's schedule, the Lovelace staff began calling potential candidates to schedule testing dates by telephone. In June the Clinic's staff tracked down Sarah Gorelick while she was in a Kansas City beauty shop having her hair done. A male testing candidate had canceled on short notice, and the Lovelace Clinic staff wanted her to fill the open slot. Her official invitation arrived in the mail after she had already left for Albuquerque.[14]

The women who received invitations to Lovelace's Woman in Space Program responded enthusiastically because they recognized what a rare chance they were being offered. In the early 1960s, women remained a minority in aviation, especially at more advanced levels. According to the Federal Aviation Agency's records, as of January 1, 1960, only 3,246 women held private pilot's licenses nationally. Of these, 782 qualified as commercial pilots. The number of women with air transport ratings was even smaller. According to the FAA figures, only 21 women qualified as airline transport pilots. Even without any assurance that Lovelace's initiative would yield a future women's astronaut program, it offered an attractive opening. With the suggestion that it might lead to something else down the road, it proved even more exciting.[15]

Among the women themselves, beliefs about whether the Lovelace examinations would produce a women's astronaut program varied. Given the tendency for later life experiences to color how earlier events get remembered, it remains difficult to judge precisely what expectations the participants held for the program's future at the time. Gene Nora Stumbough Jessen asserts that she never thought that Lovelace's testing program had any future beyond the week of testing in Albuquerque. Likewise, Virginia Holmes recalled, "I did not think it would go far and of course, it didn't." Yet for others who participated Lovelace's personal and professional connections suggested that his Woman in Space Program offered longer-term possibilities. Opening the new frontier of spaceflight meant that old prejudices might be trumped by scientific rationale supporting women's participation. Sarah Gorelick Ratley remembered being inspired by President Kennedy's inaugural call for Americans to do something for their country. Wally Funk also became very excited about the potential of Lovelace's initiative.[16]

Such speculation about a developed program for women astronauts seemed reasonable because in early 1961 no human being—Soviet or American—had ever flown in outer space. Jerri Sloan Truhill recalled that at the time her image of the U.S. space program consisted mostly of rockets exploding on launch pads. Nonetheless, when she found out about the Lovelace program she thought, "Well, they must know something I don't know. They must have something that will get up there that I haven't heard about. If they think that they can get something up there, I sure would like to fly it."[17]

Regardless of whether they thought Lovelace's program had a future, the women who responded to the offer willingly rearranged their schedules to seize the opportunity. Because she had just graduated and would not be returning to the university as a flight instructor until fall, Gene Nora Stumbough took advantage of the university's summer break. She reshuffled her summer plans, using the time to prepare herself for the Lovelace examinations before traveling to Albuquerque. To complete the examinations, Sarah Gorelick took time off from her engineering job at AT&T and Long Lines Engineering. Since she had already expended her vacation time to fly in air races, she opted to take unpaid leave. Both women found the flexibility they needed to take the tests.[18]

The women who owned their own businesses had more freedom to take the time off, but they also felt the pressures very personally. Since Pat Jetton had partners in the fixed base operation (airplane servicing station) that she and her husband owned as a part of a small group, she rearranged her work schedule. As she recalled, "I had two children, but my mother lived with me and I had a maid." Bernice "B" Steadman found the time to go to Lovelace despite owning her own flight school and being in the middle of moving into a new house. Owning businesses meant these women had to cover their own work responsibilities. However, that ownership also meant their job security would not suffer.[19]

Some of the women did not have that reassurance; they risked good aviation jobs to participate. Georgiana McConnell recalled that, despite a new job, taking the Lovelace examinations remained her priority. "I had just begun work at a new Piper Dealership when I got the word to report to the Lovelace Clinic. If there had been a conflict about time off, I would have quit." At her workplace, Irene Leverton confronted the owner of the air taxi and flight school where she worked. The Los Angeles–based operation ran air taxis for the Hollywood stars and offered flying lessons. When the Lovelace Clinic first called Leverton, the owner refused to grant her the vacation time she needed. Over his objections, she accepted Lovelace's second invitation, taking a week off from her air taxi assignments. When she returned, she had been demoted to instructing primary students.[20]

The examinations presented such an extraordinary opportunity that when other pilots who were not on Lovelace's original lists heard about the testing, they volunteered themselves. After talking to Cobb about her testing experiences, Fran Bera sent a letter to Lovelace in September 1960. A month later, in October, Wally Funk saw *Life* magazine's coverage of Cobb's second-phase testing and wrote to Dr. Shurley to volunteer. After the *Parade* magazine article featuring Cochran and the Dietrich twins appeared in April 1961, women pilots began sending letters to Cochran, volunteering to take the tests. Two of these women (Marjorie Dufton and Kay Cagle) eventually earned invitations to Albuquerque.

Reports of the Lovelace program also spread through women's aviation events in 1960 and 1961. Although rumors of a female astronaut testing program had circulated on the women's air racing circuit for some time, Bernice Steadman first took the news seriously when she talked to Jean Hixson at an air race. Hixson had already been accepted into Lovelace's program. Gene Nora Stumbough heard about Lovelace's initiative from Wally Funk at a collegiate air meet. She recalled, "So, I wrote Dr. Lovelace and I basically told him that I just didn't see how he could do this program without me." After Stumbough submitted her qualifications, Lovelace invited her to become one of the last pair of women to participate in the Woman in Space Program.[21]

Her partner, Jane Hart, learned about the Lovelace testing through a friendship based in the Ninety-Nines. After Hart met Bernice Steadman at a Ninety-Nines meeting, their friendship deepened when Steadman instructed Hart for her instrument rating. When Steadman contracted hepatitis before she was to report to Albuquerque, she and Hart went sailing together while Steadman recuperated, regaining her strength for the tests. When they returned, Steadman volunteered her friend for the project. "I said, 'Dr. Lovelace, you have got to have this Janey Hart. She has had eight kids and she is in *physical* condition. You really have got to do this.'" Lovelace accepted Hart for the last testing pair even though, at thirty-nine, she was already four years older than the cutoff age. With these last two volunteers—Stumbough and Hart—nineteen of the twenty-five women that Lovelace contacted completed the astronaut tests at the Albuquerque Clinic. They followed a schedule identical to what NASA's own astronaut candidates had undergone in 1959.[22]

"Sometimes All We Could Do to Keep Going Was to Laugh at Ourselves"

When the women pilots arrived in Albuquerque, they embarked on the most thorough physical examinations yet assembled. Although each of the women submitted to the examinations alone, they all underwent the same testing regimen. The Project Mercury examinations consisted of five days of rigorous testing conducted on a tightly run schedule that sometimes did not even permit time for meals. To ensure the astronauts' safety in the unforgiving environment of outer space, the Lovelace Clinic physicians poked, prodded, x-rayed, and analyzed every part of each candidate. They attempted to discover anything that might affect performance in space. Although the Foundation created some new tests, most of the schedule incorporated examinations widely practiced by contemporary physicians. In the Project Mercury physicals and in the Woman in Space Program, the Lovelace physicians' innovation came not from creating new testing techniques but from assembling a schedule that was revolutionary in its comprehensiveness.

Upon arriving in Albuquerque, each of the women checked into the Bird of Paradise Motel across the street from the Lovelace Clinic. A utilitarian motor inn within convenient walking distance, the Bird of Paradise served as a place to collapse at the end of the day and to administer the cleansing regimens prescribed in the tests. Each morning the women walked across the street to begin a schedule of physically draining examinations inside the Lovelace Clinic's modern pueblo-style walls. In the evenings they crossed four lanes of traffic on Gibson Boulevard to return to what some called the "no tell motel" and prepare for the next day's tests.[23]

When each candidate checked in at the Clinic on a Sunday, she received a detailed schedule and a packet of materials from the clinic operator. For every day of examinations, the agenda listed directions for medical appointments beginning at 7:30 or 8:00 in the morning. With the exception of sometimes varying the examination order to accommodate the Clinic's schedule on a particular day, the women followed exactly the same regimen that had been created for NASA. The Lovelace physicians designed the tests to elicit the most complete information possible about a human being's health.[24]

Before a candidate began any physical testing, however, she had to complete a detailed family medical history and aviation record. "Upon arriving," Georgiana McConnell recalled, "I went to the Clinic to pick up what was needed. I was given a sheaf of cards to complete with family information and personal information." Those without complete medical histories reaching back to all four grandparents had difficulty answering all the questions on the Cornell Medical History forms. In the packet of materials, each candidate also received computer punch cards to use in recording her personal aviation and medical history. In fact, for all the examinations, the Lovelace physicians used IBM computers to compile test results encoded on specially designed machine-read cards.[25]

The Lovelace physicians needed such an advanced data processing method because discerning whether candidates met the exacting standards of space fitness required immense amounts of information. Since outer space presented dangers beyond any encountered on earth, they needed to determine more than just the absence of disease. Even a highly fit candidate might have conditions that would prove dangerous during a mission. As a result, physicians sought out any physical imperfections that might be undetectable on earth but potentially fatal given the rigors of spaceflight. For example, the Lovelace examiners tested for minute openings between the right and left chambers of the heart that would be too small to cause any clinical symptoms on earth but could cause problems if the subject experienced a rapid loss of atmospheric pressure. The Lovelace physicians innovated some diagnostic apparatus to examine the candidates. More often, however, they used established methods, redefining their standards to meet the extreme rigors of outer space.[26]

As was appropriate to the most thorough physical examinations yet conducted, the week's itinerary offered detailed instructions to prepare each candidate. In addition to limiting or prohibiting eating and drinking, the schedule also restricted smoking and prescribed stool and urine samples to be taken throughout the week. Typical instructions included "nothing to eat, drink or smoke after midnight until after completion of laboratory tests on Monday morning." The schedule even prescribed hair washing on the evening before the EEG (electroencephalogram). It specified that no hair dressing of any kind, "oil, spray, or anything," be applied until after the EEG. (Both men and women received this instruction.) Each woman spent her week carefully following the appointment schedule from test to test.[27]

Although the Lovelace physicians added a gynecological checkup to the women's schedule, their testing experience so closely duplicated the Project Mercury candidates' itinerary that the women's appointment schedule did not include it. In a letter to her parents, Gene Nora Stumbough recorded that her pelvic exam occurred on the first day, after Dr. Luft's pulmonary function test.

As she wrote, "End of the day—Dr. Barber at OB x Gyn—pelvic exam. He says I'm a nice normal-type girl." Despite the fact that all of the women who underwent the Lovelace tests recall having this test, none of the surviving schedules list that stop on the day's agenda.[28]

Adding a gynecological component was the only systematic adaptation that the Lovelace physicians made to the Project Mercury physical during the Woman in Space Program. They asked a couple of the women to keep a basal temperature record, recording their body temperature for five minutes each morning before eating, drinking, or smoking, at the same time each day over the course of a month, beyond the time when they left the clinic. Such charts, commonly kept for natural birth control or promoting conception, monitored temperature changes related to ovulation. Asking active women to keep elaborate temperature records in the midst of busy flying schedules proved unrealistic, however.

Sarah Gorelick Ratley remembered asking one of the Lovelace nurses to accompany her when she bought the special metabolic thermometer because she was terribly embarrassed, as a young single woman, to be buying a thermometer specifically designed to track ovulation. She kept the chart for two weeks, then missed a week. She tried diligently to take the measurements as she had been instructed, but the process grew increasingly annoying because she was busy flying in an air race. Finally, thoroughly frustrated by the inconvenience of monitoring her basal temperature while racing airplanes, she lost her patience and threw the thermometer out the window at an air base. Because basal temperature monitoring proved so unmanageable, the Lovelace physicians dropped this component of the gynecological testing, leaving only the pelvic examination to become a standard part of the women's regimen. But plenty of other tests remained to fill out the schedule.[29]

The Project Mercury astronaut selection physicals consisted of so many individual tests that even a concise explanation of the schedule published in a journal article required almost twenty pages. After the comprehensive family health history and a detailed aviation history, the physical examinations began with a complete checkup conducted by "a physician who is both an internist and [a] flight surgeon." More than a dozen separate eye tests followed. Hearing was tested in a specially constructed soundproof room. Cardiologists tested the subject's circulation as she lay on a tilt table (to identify circulation problems that could present themselves in zero gravity). The Lovelace physicians employed supersensitive screens and "ultra fast x-ray film" to allow for the required number of x-rays without exposing the testing candidate to dangerous levels of radiation. Urinalysis and other laboratory tests monitored general electrolyte levels and blood composition. Injecting the subjects with dye and measuring its absorption tested liver function; other tests measured the percentage of body water. Physical competence tests determined cardiac strength and lung power. Additional experiments measured lung capacity.[30]

Finally, at the Los Alamos Scientific Laboratory, Lovelace's associates used a human radioactivity counter to measure total body radiation count and potassium levels. At the end of the week, each candidate flew to Los Alamos for this last test. The subject lay on a gurney-sized tray that slid into the wall-sized detecting instrument. Just in case the coffin-sized opening proved too claustrophobic, the operators gave each person a "chicken switch" to hold. A push of the button would signal the researchers to remove the panicked candidate from the machine. It would also signal the end of that subject's participation in the overall testing program. An astronaut could not be uncomfortable in close spaces.

The completeness of Lovelace's testing regimen exhausted its subjects. As Jerri Sloan Truhill recalled with a rueful smile, "We did not have any secrets when we got out of there." The other women shared her sentiments. On the night before her last day of testing, Gene Nora Stumbough summoned the energy to write a detailed letter to her parents about the examinations she had been taking since early Monday morning. "Dear folks," she wrote, "Believe me, we've been on a constant run and these are no picnic. I never knew there were so many parts of the human body to be explored—all they did was poke tubes up me and down me and take blood out of me and put shots in me." Conducting all the tests in one week meant that the examinations sometimes contradicted each other. As Stumbough recorded about the first day of testing, "After Dr. Secrest we had our first meal of the day—lunch. Then two enemas—so why eat?" The packed schedule and constant cleansing regimens challenged the women's enthusiasm. Keeping their spirits up became another test that the Woman in Space Program candidates had to pass.[31]

Since the women experienced the Lovelace testing in staggered pairs or solo, not as a unified group, each of them needed to find ways to cope with the stresses of the week. Unlike NASA's male astronaut candidates, who underwent the Project Mercury physical together, the female candidates never met as a group during the 1960s. At most, each woman could rely on her testing partner or the Lovelace staff. The women Lovelace tested used determination, humor, and camaraderie to endure the isolating experience.

Several of the women experienced the testing alone. Georgiana McConnell found herself taking the tests at the same time as a single male candidate. Without a female partner to share her experience with, McConnell relied on the Lovelace staff to help her through some uncomfortable examinations. In particular, the clinic nurses comforted her when an already difficult test dragged on much longer than expected. To examine stomach acids, the women had to swallow a length of thin hose. As McConnell recalled, "I gag easily and that was terrible. . . . Tears came into my eyes. Then they couldn't find the acid and gave me a shot to stimulate it. I had to lie there with the tube down my throat for thirty minutes before being able to test again. The nurses were so sweet because off and on tears flowed down my face while they soothed me." Sarah Gorelick had a very similar experience. Since she had filled a man's testing slot at the last minute, she found herself taking the tests solo. As a result, a nurse at the Lovelace Clinic befriended her, and in the evenings, they went out to Old Town Albuquerque together.[32]

Marion Dietrich also underwent the examinations alone, but she drew confidence from her twin sister, who had taken the tests a few weeks earlier. While she was at the Lovelace Foundation, Jan Dietrich wrote nightly letters to her twin. As a result, Marion knew what to expect from some of the more trying examinations. Jan warned Marion to take advantage of any time she could get to eat; the schedule often forbade meals or did not leave sufficient time for food. "Come with a little extra weight," she advised; "you miss one or two meals every day."

Marion also had Jan's reassuring assessment of the pain involved. Although most of the Lovelace examinations proved only bothersome in the cumulative effects of being continually poked and prodded, a few caused discomfort or even pain. To examine inner ear function, an otolaryngologist induced vertigo by squirting icy water into the ear canal and then measured how quickly the subject recovered. As a part of the neurological examinations assaying reflexes, one painful test stimulated the ulnar nerve with electricity by inserting a large "coaxial needle electrode" into the arm. Each candidate also endured a barium enema for fluoroscopy imaging. Given the fatal risk of sending an unfit astronaut into outer space, the Lovelace physicians willingly sacrificed candidates' comfort for attention to detail. As Jan informed her sister, however, "Under normal circumstances, pain often means injury—and this is probably what is upsetting to most people. But at Lovelace, at no time do they do anything that could even possibly injure you. Thus, even if something is momentarily annoying, it is no real problem." When Marion took the tests she persevered, knowing that the Lovelace staff would not hurt her—and that her twin sister had already withstood these same examinations.[33]

Even if the women undergoing the examinations had a testing partner, the pairs did not attend clinic appointments together. During the day, each woman had her own schedule and moved from department to department alone. As Gene Nora Stumbough Jessen recalled, "We didn't do it together. We'd start the day going different places, to different tests. And end up at the motel in the evening. So it's not like we went hand-in-hand to all those tests." Because they were not together much during the days, the women who had testing partners talked when they returned to the motel. Jessen remembered "comparing notes in the evening" as a way of offering each other support.[34]

Other participants also found resilience in the camaraderie of going through the examinations paired with another candidate. Joan Ann Meriam and Pat Jetton spent their week in Albuquerque together. As Jetton recalled, that meant getting to share some of the more awkward moments with someone else. In one instance, she recalled, their plans for a nice meal ended abruptly. "Joan and I went out to dinner one evening after testing at a restaurant some distance from our hotel. We had been given some bottles of orange juice, by Lovelace, to drink earlier and had done so. We had just been served our meals when it became apparent that the juice had also contained some mineral oil. We jumped up, quickly drove back to the hotel and barely made it back in time." They had learned their lesson: "[The] next time we were given something to drink we found out what it contained before making our dinner plans."[35]

In many cases, humor helped the women to weather examinations that turned their stomachs, threw them off balance, and strained their modesty. Jane Hart and Gene Nora Stumbough sat together in the motel at night laughing about the day's indignities and warning each other about upcoming tests. As Stumbough reported to her parents in a letter from Albuquerque, "She has a tremendous sense of humor and sometimes all we could do to keep going was to laugh at ourselves." Quick friendships formed under the duress of the long days of testing. Both Jerri Sloan Truhill and Bernice Steadman recalled that their partnership helped them to prevail over physical ordeals. As Steadman remembered, "It was an interesting experience to do and it was even made more fun in being able to share it, Jerri and I. We've laughed about these things. If we had been going through it all by ourselves, I think it would not have been nice at all." In the evenings, they sat on the porch of the motel having what they called their "Lovelace cocktail hour," drinking concoctions designed to clear their systems for the next day's tests.[36]

Whether going through in a pair or by herself, however, each woman had to find a way to endure when left alone. Since Steadman had just moved to a new house on a farm, she found a unique way to keep her mind occupied. "So as a part of staying awake [during a test], I arranged all of the furniture in my mind. So when I got home I had an easy job. On the phone [in the evenings] I told Bob where to plant the roses and stuff." The mental gymnastics helped her stay alert.[37]

Whether or not the women had testing partners, the Lovelace staff encouraged them. Gene Nora Stumbough Jessen recalled having a general feeling that the doctors there wanted the women to do well (especially one of the physicians conducting an eye test). Wally Funk, the youngest woman to go through the tests, remembered Dr. Lovelace as wonderfully helpful, "a great big daddy" who wanted the women to succeed. For their part, the Lovelace physicians also noticed the talents and determination the women exhibited. As Dr. Donald E. Kilgore, one of the Lovelace physicians, recalled, "They were awesomely intelligent. They were incredibly motivated." In the end, it showed in their test results.[38]

Results . . . and More Tests

Of the nineteen women who took part in Lovelace's Woman in Space Program in 1961, twelve succeeded. Including Jerrie Cobb, thirteen women passed the same physical tests NASA used to select its astronauts. Since the women took the Project Mercury physical over a period of months, Dr. Lovelace informed each woman of her results as she finished. In the process, he compiled a list of women for whom he planned a further phase of advanced aeromedical testing.

Approximately one month after she returned to Dallas from her early April tests, Jerri Sloan received a letter from Lovelace that explained his hopes for the group's future and her place in them. Dr. Lovelace informed her that he was making plans for the successful candidates to meet as a group for additional testing sometime in the summer. He wrote, "By sometime in June we hope that all of you that passed the examinations here will be able to go in a group to a service laboratory where further test procedures will be carried out. Just as soon as a definite date is picked, I will let you know immediately." He suggested that she concentrate on getting in shape, "as the forthcoming tests are going to require considerable physical stamina." To achieve this, he prescribed "walking, swimming, and bicycle riding as well as calisthenics." Almost as an afterthought, he notified Sloan of her success in the preceding tests. "I am happy to say that you were one of those that were successful in passing the examination," he closed.[39]

Sloan had a good idea how well she had performed from the assessment given at her exit interview at the Lovelace Clinic. On the last day of the schedule, each candidate met with Dr. Lovelace or another doctor to discuss the week's data. Even though some test results had not yet been processed, the physicians could offer a fairly accurate picture of the outcomes. The final evaluations arrived later by letter.

Rejections took several forms. Virginia Holmes received a letter thanking her for her participation in the Lovelace examinations but without any invitation to continue. The notice never said she had failed; it simply did not extend the invitation to participate further. The result did not surprise her. During the interviews, she had told the physicians about her significant work and family responsibilities. In addition, Holmes felt certain she had disqualified herself during the human radioactivity counter test. Although she did not push the "chicken switch" to escape from the tomblike testing chamber, she knew the physicians probably detected her sudden onset of claustrophobia from her racing heart rate. Georgiana McConnell learned about her test results in a much more direct letter. It read, "We regret very much to inform you that you did not meet the requirements for the women astronauts program during your recent examinations here." Neither she nor five others advanced to the program's next stage.[40]

The Lovelace examinations established such stringent standards that women who did not pass nonetheless were in very good health. In fact, the exacting doctors sometimes detected conditions, even in the women who passed the tests, that normal examinations never duplicated. Gene Nora Stumbough Jessen remembered that the cardiologist found a "systolic heart murmur" after he had the fluorescent lights in the room turned off because their humming interfered with his examination. No doctor had detected it since. When Fran Bera met with the physician at the end of her week in Albuquerque, he informed her that she had some "unusual brainwaves" that probably would never pose a problem to her health. She had herself rechecked at two hospitals, but the aberrant patterns could not be detected again. Bera continued to fly successfully, winning the All-Woman Transcontinental Air Races multiple times.[41]

The Lovelace physicians also detected a "brain abnormality" in Pat Jetton that ended her participation in the Lovelace testing but never affected her afterward. Jetton concluded her Albuquerque visit by setting an approximate date to attend additional testing but returned home to find a letter stating that a brain abnormality would prevent further involvement in the program. She recalled, "My reaction was total fear—I was convinced that I was going to have a stroke or other problem. I worried for some time before finally going to a doctor who was totally unconcerned about the report." Regardless, she had been disqualified from the next phase of examinations.[42]

Like the multistage process that NASA used to select the original Mercury astronauts, Lovelace's Woman in Space Program also had different phases. First, physical examinations eliminated any candidates with symptoms indicating potential problems. Next, psychological evaluations tested for mental stability and desirable personality traits. These examinations, which some called "Phase II," included the sensory deprivation experiments that Wally Funk had seen Cobb completing in *Life* magazine. Finally, sophisticated diagnostic and simulation technologies, including jet aircraft, tested the candidates' physical performance under extreme g forces.

The Lovelace Foundation's physical examinations constituted only one part of the process that NASA used to select its astronauts. After passing the Lovelace tests, the male astronaut candidates went from military base to military base in 1959, testing their ability to adjust to high-altitude conditions, water bail-outs, and extreme g forces. When Lovelace wanted the women he had recruited to take the same kinds of tests, however, their gender precluded use of the necessary facilities. As a result, he set up informal arrangements to duplicate the original astronaut fitness examinations as closely as possible.

Only a few women completed more than one portion of the testing. As Lovelace narrowed the group from twenty-five invitees to thirteen successful candidates, some of the women took further examinations. Because of time constraints on the use of equipment, however, Lovelace did not ask all of those who passed the physical component to take the psychological testing in Oklahoma City.

Some women had difficulty freeing themselves from work or other personal responsibilities in order to take more tests immediately. As a result, Lovelace did not require them to take the "Phase II" tests. Jerri Sloan did not attend because of her child care responsibilities. Gene Nora Stumbough and Jane Hart went through the Lovelace Foundation tests so late in the summer that there was no time to go to Oklahoma City before attending other testing that Lovelace planned. (Only Hixson finished later than they did.) Although Lovelace wanted more information about each of the candidates, the loosely organized nature of the Woman in Space Program did not require that each of the women go through all of the steps in order.[43]

In the end, only three women—Jerrie Cobb, Wally Funk, and Rhea Hurrle (Woltman)—completed the psychological examinations. Toward the end of the summer of 1961, Hurrle and Funk each traveled to Oklahoma City to take Dr. Shurley's isolation tests. As the women took batteries of other psychological tests, they also prepared for these experiments. When each candidate entered the room with the sensory deprivation pool, the temperature of the air and the water matched her body temperature. As a result, when the subject floated in the pool, she could not feel the water or air on her skin. Wally Funk remembered trying to pound on the water's surface with her hands and being amazed that she could not feel the water splashing on her. The researchers also plugged the subject's ears and blocked all light. For as long as she could, each participant floated in darkness, buoyed by a few small pieces of foam rubber.[44]

Hurrle and Funk completed Shurley's tests in close succession. Hurrle spent time during the last week of July 1961 floating in the pitch-black pool. Her tests concluded on Wednesday afternoon, the same day Funk arrived. Funk underwent the psychological and psychiatric tests throughout the first week of August. Since Cobb lived in Oklahoma City, she arranged for Funk to visit her during her stay in town. Other than this social contact, however, each woman underwent the second-phase testing regimen alone. Both performed well. In fact, Funk floated so calmly and quietly in the sensory deprivation tank that the staff eventually called her out of the water after ten hours and thirty-five minutes.[45]

Despite Lovelace's having only three women undergo the psychological simulations, this phase of the Woman in Space Program also reinforced the impression of an organized project. When Funk had originally volunteered for the Lovelace testing by writing to Dr. Shurley about Cobb's sensory deprivation testing in October 1960, he politely rebuffed her by referring her to Lovelace or Cobb. After Funk participated in the Oklahoma City testing, however, his correspondence differed significantly. Unlike his earlier letter, in which he stressed that his participation was "on a purely informal basis," this time he concluded by referring to his part in Lovelace's initiative as "this segment of the selection program." The perception of Lovelace's project as the precursor to organized recruitment of women astronauts may have contributed to Funk's dedication to finishing all three phases of testing.[46]

Technically, only Cobb completed the third phase under Lovelace's auspices. In early 1961, Cobb underwent additional tests at the Naval School of Aviation Medicine in Pensacola, Florida, examinations that included high-altitude and g-force testing. When that opportunity was closed to the rest of the group, however, Funk later took similar tests on her own initiative. As she recalled, "I just wrote letters and asked, 'I'd like to be a subject. Would you allow a civilian [to take these tests]?" Through one of her flying students, Funk arranged to take physiological examinations at El Toro Marine Corps Air Station in California. At El Toro, she took both an altitude chamber test that simulated high-altitude flight and the "Martin-Baker" ejection seat tests, designed to mimic ejection from an aircraft. Funk also accumulated testing experience at another site. Again through personal arrangements, she arranged to take a ride in a human centrifuge at the University of Southern California.[47]

Although Funk managed to negotiate a ride in the human centrifuge, she could not obtain the crucial piece of equipment she needed to withstand the test successfully: a partial pressure suit. Because she knew that, as a civilian and a woman, she would not be issued a PPS, she constructed a makeshift one, unknown to the testers. Before the centrifuge experiments, Funk called her mother and asked to borrow her "worst merry widow and a couple of girdles." From these pieces of restrictive feminine underwear, Funk created a tight-fitting garment encasing her pelvis and lower torso. She amazed the researchers with her ability to withstand greatly increased gravitational forces, supposedly without a pressure suit. By taking additional tests on her own, Funk found a way to circumvent women's limited access to high-tech aerospace simulations.[48]

The way Lovelace arranged the various testing opportunities for women in the early 1960s demonstrated how formidable were the institutional and financial barriers blocking women from space experimentation. When Funk initially applied on her own, Shurley rebuffed her. Without someone to vouch for her and pay for the examinations, she could not participate. With the Lovelace Foundation backing them in the Woman in Space Program, however, female pilots could undergo both the Project Mercury physical and advanced psychological experiments. During the summer of 1961, as he accumulated a list of women who had passed his Clinic's astronaut selection physical, Lovelace once again arranged to circumvent the barriers to advanced space testing.

Preparing for Pensacola

With Lovelace's help, Cobb went further than any woman had gone before in being tested for space fitness. She had exhausted the tests at Lovelace's Albuquerque facilities. She needed advanced aeromedical equipment. In February 1961 Cobb contacted the air force laboratories at the Wright Air Development Division about her undergoing stress tests there. Although they refused to let her use their equipment, Lovelace and Cobb arranged with military researchers for the U.S. Navy to permit her to undergo tests at the School of Aviation Medicine in Pensacola, Florida. In the spring of 1961, while the other women hopefuls traveled to Albuquerque to test their mettle at the Lovelace Clinic, Cobb arrived at Pensacola. The correspondence involved in organizing this one-time exception revealed the reluctance military officials still felt toward women's astronaut testing.

When Cobb's participation required official approval, the military establishment responded with jokes. In order for her to fly in navy jet aircraft for an airborne EEG, a test that measured brainwaves as the subject underwent various g loads in a jet, the Naval School of Aviation Medicine needed approval from navy headquarters. On receiving Lovelace's request, the Pensacola base wired the Pentagon requesting permission to use navy materiel to determine the difference between male and female astronauts. Unable to pass up a straight line like that, the Pentagon replied: "If you don't know the difference already, we refuse to put money into the project." Those in positions of power did not take women seriously as potential astronauts.[49]

Despite being the subject of jokes, Cobb took the Pensacola testing quite seriously, demonstrating that women could succeed in the simulations. Her May 1961 performance paved the way for Lovelace to begin planning the same opportunity for the other successful female candidates. While women continued to make the trip to Albuquerque to be examined, Lovelace made informal arrangements with the navy to test the entire group of final candidates selected in the Woman in Space Program.

As soon as Cobb found out that Lovelace had made an agreement with the Naval School of Aviation Medicine, she wrote to notify the women who had already passed the Lovelace tests of this extraordinary opportunity. To include the whole group, she coined the acronym FLAT, addressing her note "Dear F.L.A.T. (fellow lady astronaut trainee)." She informed them that beginning on July 16 and lasting until July 29, she and seven other "lady astronaut trainees" who had passed the Lovelace astronaut tests would spend two weeks undergoing intensive space simulation testing in Florida. The schedule included general physical fitness tests and clinical examinations. Endurance, acceleration, and altitude tests filled out the agenda. Most notably, however, the navy facilities offered the women the chance to fly in a jet during an airborne EEG.[50]

With the announcement of the Pensacola examinations, the Woman in Space Program took on the added pressure of military involvement. As the self-appointed leader of the group, Cobb took it upon herself to preemptively insulate the navy from any risk in testing women. She sent each of the women a release form declaring that she was participating entirely on her "own initiative, risk, and responsibility" and absolving the U.S. government of any responsibility for her injury or death. The candidates received instructions to have someone witness their signatures before sending the forms back to Cobb's home.

Cobb also warned the women not to initiate any press coverage or draw any attention to these plans. Beyond paperwork and media restrictions, the women's preparations mostly involved physical conditioning. Cobb urged her fellow pilots to get into the best physical condition possible to improve their chances in the coming tests. Each on her own, the women began preparing for their Florida tests.[51]

Plans for the Pensacola tests hit a setback, however, when Lovelace postponed their scheduled date. Ten days before the original test date arrived, Lovelace sent a letter notifying the candidates, "Originally these tests were set for July 18. It has been necessary to change the testing date to begin on Monday, September 18." Lovelace did not offer any explanation about why the session had been postponed. Other potential candidates continued to undergo the initial examinations at the Lovelace Clinic, and he may have wanted to wait for a complete group before taking advantage of what would surely be a one-time use of the navy facilities. Or navy officials may have rescheduled. Regardless, Lovelace informed the test candidates that he continued to work on the details and would send more information soon.[52]

In the letters informing each of the newly successful candidates about the Pensacola tests, both Cobb and Lovelace had emphasized that the women should get in shape for the tests. As a result, from the first notice of the Pensacola plans in early June, the women trained. They bicycled, walked, swam, and hiked. Each tried to plan her conditioning so that she would arrive at Pensacola in peak form in July. Now, with the postponement notice, they extended their exercise regimens, setting their sights on September.

Despite the postponement, Lovelace's letter solidifying the plans for the new September dates offered even more encouragement to women who dared to hope that the Woman in Space Program might become an official female astronaut project. Lovelace informed the group about Cochran's sponsorship of the Pensacola arrangements. Along with the details of the world-famous pilot's planned attendance in Florida, however, Lovelace also offered specific suggestions for how the successful candidates should handle any publicity that might eventually occur. He advised the women, "Immediately after the final selection is made of those of you that pass the tests at Pensacola it is suggested that you have a meeting as a group to decide on a group policy for any publicity. As you know, the male astronauts have acted as a group on all matters concerning publicity every [*sic*] since their initial selection and I would like to strongly urge that the results of their group acting in this field be considered very seriously." Given the attention that Cobb had received after Lovelace first announced her successful results, his advice made sense.[53]

Having Lovelace compare the women's group to the Mercury astronauts carried extra weight, however, given its source. As the chairman of NASA's Life Sciences Committee and the head of the Mercury astronauts' physical selection, Lovelace spoke with firsthand authority. Also, Lovelace's words suggested that the women might be facing not a brief spate of media attention but extended coverage. By asking them to consider the male astronauts' plans, he suggested that they needed more than a simple strategy for one press conference or even a week's attention. He implied that this new group could expect ongoing media scrutiny. Finally, Lovelace compared the end of the Pensacola phase with the Mercury astronauts' "initial selection." Reading those words coming from the man who helped select NASA's own astronauts could not help but suggest that the Woman in Space Program might have a future beyond Pensacola.

In his advice about public relations, however, Lovelace never demanded absolute secrecy. Although many assertions have been made about the secrecy of Lovelace's Woman in Space Program, beyond the low profile requested of each candidate regarding her initial testing in Albuquerque or cautions about "individual publicity," the program as a whole never remained a secret. After Lovelace's Stockholm announcement in August 1960, the women's testing never was a secret. Nor was news of Lovelace's women's astronaut testing program suppressed later. In fact, the press regularly reported the story in national media. In January 1961 an Associated Press story announced the existence of twelve women scheduled to undergo physical testing at the Lovelace Foundation. While the examinations continued in Albuquerque, the April 1961 *Parade* magazine cover article (featuring Cochran and the Dietrich twins) appeared in newspapers across the country. Finally, as the successful candidates prepared to go to Pensacola during the summer of 1961, Marion Dietrich sold the story of her participation to *McCall's* magazine.[54]

Dietrich's article, "First Woman in Space," appeared in the September 1961 issue. A photograph caption explained, "Five American women—two of them twins—have passed the preliminary tests for astronaut. Here is the exclusive-to-McCall's story by one of the twins—Marion Dietrich." Dietrich described herself as a "fairly typical girl," interested in clothes and landscape architecture, playing the piano, and going to parties. She also recounted her early interest in aviation, spurred by her father's World War I piloting, and her own blossoming career in aviation. Currently, she explained, she worked at an airport in addition to writing for magazines and newspapers.

Dietrich's account of receiving her invitation from Dr. Lovelace emphasized how excited she was to receive a letter from someone with his impeccable credentials. In the article's introduction, Dietrich recalled the thrill of getting a letter asking, "Will you volunteer for the initial examinations for woman astronaut candidates?" As she wrote for *McCall's* readers, "In stunned disbelief, I reread the letter's first paragraph and began the second. . . . My glance darted to the signature. Dr. W. Randolph Lovelace II, Director of the Lovelace Foundation. That is the astronaut physical-testing center, in Albuquerque."

Dietrich's article did not mention any testing beyond the first stage, but its conclusion demonstrated the author's interest in pursuing the dream of spaceflight: "If and when the second miraculous question is posed: 'Will you volunteer for the space flight?,' I won't be able to answer fast enough. I'm ready! How soon do I go?" Her article publicized Lovelace's testing just weeks before the scheduled Pensacola trip.[55]

For some of the participants, postponing the Pensacola testing dramatically increased the personal costs of participating. Taking the initial examinations had required only one week, albeit often on short notice. With the Pensacola tests on the agenda, the women needed two full weeks free. Several unsympathetic employers proved reluctant to grant more leave time. As a result, as the women prepared for the September testing dates, some of them faced difficult choices between keeping a job and pursuing astronaut testing.

To support Jerri Sloan's request to spend significant time away from her job,

Jerrie Cobb wrote an appeal on her behalf. Cobb pulled out all the stops. Her letter mentioned her own consulting with NASA, implying that NASA administrator James Webb's appointing of Cobb as a space agency consultant had been an endorsement of women in space. Rather than simply appealing to the power of Lovelace's private affiliations, Cobb used Lovelace's NASA title and mentioned Flickinger's air force rank. Furthermore, she wrote, "This serious program is being conducted on a highly scientific level and is of utmost importance. The Pentagon has recently approved the further testing of the women pilots who successfully passed the Albuquerque tests, at a military Aerospace medical facility for the last two weeks in September." Cobb touted the program's scientific importance, current secrecy, and coming publicity. "Although it has been necessary to keep this program 'under wraps' as much as possible, after the September tests, the results and names and details will be released. Life Magazine, among other news media will carry the stories." Whether or not her boss believed Cobb's inflated descriptions, Sloan was able to make the arrangements she needed in order to attend the Pensacola testing. She avoided being forced to make a hard decision right then.[56]

As Sarah Gorelick considered taking the Pensacola tests, however, she could not dodge the difficult choice facing her. She would not be allowed to take the time off. Early in 1961, she had scheduled her vacation from AT&T's engineering department so that she could fly in the Powder Puff Derby. Because she had already used her vacation time when she received Lovelace's invitation, she took unpaid leave for the week of Albuquerque testing. In July, Gorelick applied to take time off for the Oklahoma City psychological tests, only to find out too late that the examinations could not be conducted owing to equipment problems. The negotiations put her in poor standing with her bosses. In August the problems worsened when Gorelick requested more time off for Pensacola. When she realized that she would have to quit her job in order to participate, Cobb offered to intervene by modifying a copy of her earlier letter.[57]

But Gorelick's employers had run out of sympathy for her outside interests. The entire engineering department already worked overtime; they would not grant any more time off, paid or unpaid. Left with no other choice, Gorelick resigned. On September 7, 1961, her coworkers threw her a going-away party complete with a model rocket and a space helmet with "S. Gorelick" above the visor. Gorelick had thrown in her lot with the Woman in Space Program.[58]

Other women also faced scheduling conflicts because the postponed testing dates conflicted with the beginning of the school year. Myrtle "Kay" Cagle planned to begin college that fall as a first-year student at Mercer University. In mid-July, she met with the dean and requested to be excused from classes during the Pensacola testing. As evidence of the program's seriousness, she showed him a letter from Cochran advising the women on how to prepare for the Pensacola examinations. Mercer officials agreed that, in her absence, Cagle's husband Walt could register her for her courses. She would begin attending classes when she returned. As a student, Cagle could rearrange her schedule enough to permit the time off.[59]

Since she was a college instructor, however, Gene Nora Stumbough's school schedule was less flexible. To put herself through college, Stumbough taught flying at the University of Oklahoma. She had received her degree during the summer and planned to begin teaching aviation full time. As she recalled, "I went to my boss back at the University and I said, 'I've been invited to take this astronaut testing. I need two weeks off.'" Because he did not feel he could afford to lose her for the first two weeks of the term, he let her go and hired a new instructor to replace her. She felt sad to lose a position at a place where she had enjoyed her work so much, but she turned her attention to the upcoming Pensacola examinations. The thirteen women who planned to participate in the third and most advanced phase of Lovelace's Woman in Space Program had a lot riding on the tests.[60]

Whether or not they believed that Lovelace's project had the potential to turn into an official women's astronaut program—or even just an official female astronaut testing program—Lovelace's invitations excited thirteen women enough for them to risk good jobs and rearrange their personal lives to respond to his offer. In an era when women found themselves systematically barred from the newest aerospace medical technology, Lovelace's Woman in Space Program broke down barriers. With Jacqueline Cochran's financial backing, the Lovelace Clinic opened its doors to a select group of women pilots. When Lovelace exhausted his own facilities, however, he needed access to advanced equipment. Persuading colleagues and friends to volunteer their services allowed him to extend the program. Eventually, however, this system of professional favors reached its limits. Informal arrangements proved vulnerable to postponement and cancellation.

Although Lovelace had arranged Cobb's access to advanced facilities without much notice, scheduling thirteen women to take sophisticated aeromedical examinations using military jet aircraft attracted attention. The scale of the program and the media attention being paid to it made it hard to do quietly. Notwithstanding the warnings for the participants to avoid "individual publicity," stories about the never-secret program appeared in major newspapers and national publications. Marion Dietrich's *McCall's* article appeared only weeks before the Pensacola tests. Lovelace's informal arrangements began to receive formal scrutiny. Decision makers who had frowned on wasting aerospace resources on testing women in 1959 proved no more enamored of the idea in 1961.

Unaware of these developments, Lovelace continued planning the Pensacola trip. On August 21, he sent a letter to each of the women enclosing checks to pay for their transportation and their "maintenance while there, as furnished by Miss Cochran." The women made travel reservations, paying for them with Cochran's assistance. They arranged for child care and covered their work responsibilities. Those who could not rearrange their work quit their jobs. Then, on September 12, 1961, five days before they were scheduled to arrive in Pensacola, each of the women received a telegram. Its simple message quashed any hopes they may have had. The entire text read:

> Regret to advise arrangements at Pensacola cancelled Probably will not be possible to carry out this part of program You may return expense advance allotment to Lovelace Foundation c/o me Letter will advise of additional developments when matter cleared further= W Randolph Lovelace II MD

The Woman in Space Program was over.[61]

6

Jerrie Cobb, NASA, and the Pensacola Cancellation

In Pensacola, Florida, Jerrie Cobb heard the news while she was preparing for the rest of the women to arrive at the Naval School of Aviation Medicine. She immediately flew to Washington, DC, to discover what had happened. As she began making phone calls and knocking on doors, she called her Oklahoma City office and asked that telegrams be sent to her "fellow lady astronaut trainees" warning them that the trip had been canceled indefinitely. Bonnie Doyle of the Aero Commander Division of Rockwell-Standard wired each of the women: "Miss Cobb has just informed me from Washington that she had been unable to reverse decision postponing Florida testing again." The navy had withdrawn its cooperation from the next phase of Lovelace's Woman in Space Program.[1]

The advanced examinations had been postponed before, but this cancellation had the potential to be final. Since the pilots had already exhausted the examinations at the Lovelace Foundation, they needed access to more sophisticated facilities. Because most such laboratories were military, continuing the project would require government involvement.

But for a number of reasons, government cooperation had become far less likely by late 1961 than it had been when Lovelace and Flickinger conceived the women's testing experiment in 1959 or when Lovelace ran the Woman in Space Program in 1960 and 1961. Throughout the beginning of 1961, the U.S. space effort underwent tremendous changes. In the spring months while Lovelace invited female candidates to spend a week in Albuquerque,

American space policy makers had fundamentally transformed the nation's goals in the space race. And while talented women pilots drank their "Lovelace cocktails" and steeled themselves for another day's examinations, NASA had streamlined its operations toward a new national space goal: putting a man on the moon.

"Dramatic Accomplishments in Space"

The U.S. space agency began 1961 without a director and with no clear direction. The new Kennedy administration could not decide what mix of technical knowledge and bureaucratic skills an ideal candidate for NASA's top job should possess. Even after the administration had a clear job description, however, well-publicized internal disputes made the space agency's leadership post undesirable. While NASA's scientists pushed for significant advances in space science, the agency's politicians vied to match Soviet space achievements on a limited budget. In the end, James E. Webb (President Truman's former budget director and undersecretary of state), finally accepted the position because President John F. Kennedy personally convinced him that political acumen would be more important than technical knowledge. Kennedy persuaded Webb to look beyond the bureaucratic infighting to see the national and international issues involved in space exploration. Although he had found an astute political mind to guide NASA, the president still did not have a definitive space policy to offer his new administrator. Webb's new duties included defining NASA's role in the space race.[2]

Within months, the new NASA administrator found himself under pressure to commit the United States to a definitive long-term space goal because in April and May 1961, two closely spaced cold war defeats refocused the president's attention on space. First, the United States lost the race for human spaceflight when Yuri Gagarin orbited Earth on April 12, 1961. Then, only three days later, Kennedy faced the humiliating Bay of Pigs invasion. The U.S. space program responded by launching NASA astronaut Alan Shepard on May 5. Yet Shepard's Mercury capsule weighed only 20 percent of what the Soviet program had launched in Gagarin's orbit. And Shepard's trajectory was suborbital—a parabolic shot into space and then down into the Atlantic Ocean just off the Florida coast. The United States still lagged far behind the Soviet Union in heavy lifting capacity and the ability to orbit. Within days of the Bay of Pigs incident, however, Kennedy had already decided that the United States could use space as the arena to regain international prestige. Kennedy wrote to Vice President Lyndon Baines Johnson, the head of the National Aeronautics and Space Council, requesting ideas for how the United States could demonstrate American strength to the world.[3]

The famous exchange of memos between President Kennedy and Vice President Johnson redefined U.S. space policy for a decade. Their debate about the direction of the space program produced the lunar landing decision and illustrated how national decision makers weighed priorities when defining space policy. Inquiring about opportunities for beating the Soviets in space, Kennedy summarized a series of possible plans for the space effort by asking, "Is there any other space program which promises *dramatic* results in which we could win?" The emphasis on "dramatic" achievements rather than scientific or technological ones highlighted how consciously Kennedy conceived of the space program operating in the international theater of the cold war.[4]

Johnson's reply confirmed Kennedy's understanding of the connections between the space program, cold war power, and international prestige. Significantly, Johnson also focused on "dramatic" space achievements. He argued that the United States needed to "recognize that other nations, regardless of their appreciation of our idealistic values, will tend to align themselves with the country which they believe will be the world leader—the winner in the long run. Dramatic accomplishments in space are being increasingly identified as a major indicator of world leadership." He agreed with Kennedy's judgment that the United States should pursue goals that would be internationally impressive. The exchange of memos inspired Kennedy's famous declaration of a lunar landing goal.[5]

Only twenty days after Shepard's flight gave the American space program fifteen minutes of spaceflight experience—and months before the U.S. program achieved even a single Earth orbit—Kennedy stood before a joint session of Congress declaring, "I believe that this nation should commit itself to achieving the goal, before this decade is out, of landing a man on the moon and returning him safely to earth." The U.S. space program had a new mandate. James Webb's willingness to recommend a lunar landing as a possible option carried a lot of weight with key decision makers. As Senate Space Committee Chair Robert S. Kerr stated, "If Jim Webb says we can land a man on the moon and bring him back safely, then it can be done." That confidence allowed the U.S. space program to make a tremendous leap. Far from being a scientifically significant achievement, however, the moon landing received approval because, if achieved, such a feat could go a long way toward recovering international prestige.[6]

In the early 1960s, the Kennedy administration evaluated foreign policy options by assessing whether they provided appropriately strong responses to the Soviet Union. As a result, launching a female astronaut never received consideration as a worthy space achievement. Sending a woman to do a man's job would not project the image of international strength that Kennedy desired. Despite the publicity that Cobb's achievements received, having a female astronaut never made the lists of options debated by the president and vice president. In the minds of key decision makers, allowing women to participate in the dangerous pursuit of space firsts would leave the United States vulnerable to criticism, not provide a "dramatic" answer to Soviet space achievements.[7]

Within the space agency itself, NASA officials and astronauts also failed to consider launching a woman. Although Eisenhower had conscientiously crafted a space agency that would be civilian in nature, NASA's staff nonetheless consisted largely of former military men. The engineers and pilots who founded NASA in the late 1950s brought with them their military culture and its embedded assumptions about women's and men's proper roles. As such, the military ethos of the young civilian agency did not welcome women in the dangerous role of astronauts. On a practical level, once the space agency successfully launched a man into space, any physiological advantages that women offered lost their value. NASA officials could not conceive of women in the risky position of astronaut, no matter what qualifications female pilots brought to the job.

Opinions against female astronauts also ran strong among several of the Project Mercury astronauts. Within the manned space program, an often-repeated joke about female space travelers circulated: If a woman ever flew in outer space, it would be because the astronauts had been allowed 120 pounds of recreational equipment. Several of the original Mercury Seven astronauts continued to echo the disdain for women astronauts thirty years later. In their 1994 book *Moon Shot: The Inside Story of America's Race to the Moon*, Deke Slayton and Alan Shepard end their description of an astronaut's requirements (height, weight, flight time, and so on) with a dismissive "and, of course, no women, thank you." In the views of the men running NASA, women had no place in an astronaut corps newly dedicated to putting a man on the moon.[8]

In addition to focusing national space policy on the drama of a moon landing, Kennedy's lunar directive had a second consequence. The decision streamlined the space agency's agenda and ended the period of institutional flexibility in which Lovelace's program had been founded. To achieve a moon shot, the space agency concentrated its efforts. With a projected $40 billion budget to manage and a presidential directive to achieve in only nine years, NASA officials became unwilling to consider new goals that did not contribute directly to the success of the Apollo Program. After May 1961, the space agency did not want any distractions. As a result, when Cobb persisted in raising the issue of female astronauts, the NASA administrator made an unsuccessful attempt to co-opt her.

"A Great Asset to Any Part of the Program"

As the first woman to have passed the astronaut fitness tests and therefore, according to the press, America's first female astronaut candidate, Cobb became the leading advocate for the cause of women in space. As the space agency worked to respond to Gagarin's flight, Cobb brought the issue of women astronauts to its attention. In May 1961 Cobb contacted the NASA administrator, James Webb, about her interest in resuming the canceled Pensacola examinations. She also accepted a significant invitation to speak publicly on the issue to a group of experts.

On May 1, 1961, Cobb sat on a Missile/Space panel at the Aviation/Space Writers Association annual meeting. Alongside her on the platform sat various aerospace experts: a representative of Dyna-Soar (the air force's human spaceflight program); an expert on Soviet space capabilities; a scientist from General Electric's Space Science Laboratory; and George Low, NASA's director of spacecraft and flight missions. The meeting's biographical summaries listed Cobb as "the first lady Astronaut." At this meeting and news conference, Cobb presented a short paper describing the "sixteen highly qualified and experienced women pilots undergoing the Astronaut examinations at the Lovelace Foundation in Albuquerque." After explaining some of women's physiological advantages for space flight, Cobb cautioned that no women's astronaut program existed at the time and that NASA had not had any involvement in the project. Nonetheless, she urged that the United States consider "selecting and testing women pilots now for future space assignments." Although the Soviet Union had already launched the first man in orbit, Cobb hoped that the United States would achieve the first woman in space.[9]

At the end of her scheduled talk, Cobb closed by praying for the success of the next day's launch of Alan Shepard. As a devout Christian, Cobb responded to the inherently risky endeavor: "In closing, I would like to offer a fervent prayer for that kindred spirit who will be in the capsule tomorrow—God bless you—God be with you—God's will be done." She offered her prayer with the same sincerity of purpose with which she pursued her goal of flying in space. Given that NASA's George Low was sitting on the dais with her, Cobb knew that a space agency official had heard her.[10]

Nonetheless, she continued to send letters to NASA informing the agency's officials of her progress. In the midst of NASA's lunar decision, Cobb reminded the NASA administrator of her interest in becoming America's first woman astronaut. When he received her letter, Webb had just finished advising the president on the lunar launch decision, due to be announced at the end of the month. Cobb wrote that she had just completed stress tests at the Naval School of Aviation Medicine in Pensacola, Florida. Later in the spring, she had an appointment at the Johnsville, Pennsylvania, human centrifuge (a massive apparatus that generated g forces by spinning its occupant at the end of a long arm). "When that is completed I will have passed all the tests given [to] the Mercury Astronauts." She hoped to be ready when and if the space agency decided to fly a woman in space. "I have chosen to do this on my own in the eventuality that you might need a qualified woman in the space program in the future."[11]

Not only had Cobb made significant progress on completing a full set of astronaut tests, but with each step she also proved herself more determined to succeed in her ultimate goal of qualifying as a potential astronaut candidate. In her letter to Webb, she expressed her fears that the Soviet Union might beat the United States to launching a female space traveler, an accomplishment that, in her assessment, offered "tremendous prestige value." Concerned that the space agency might not be aware of potential Soviet space plans regarding women, Cobb enclosed a newspaper clipping that speculated about the launch of a female cosmonaut. Cobb concluded her message by reasserting her commitment to lobbying publicly for female astronauts. After completing her testing in Pensacola, she informed Webb, she planned to attend a NASA meeting in Tulsa, Oklahoma. Then she would fly to Paris to participate in the famous Paris Air Show, the Salon Aeronautique. Knowing that both events would draw influential figures in aviation and aerospace, she planned to use each opportunity to advocate for testing women as potential astronauts. Cobb sent the letter hoping for some encouragement from Webb; she could not have predicted his response.

The astronaut hopeful and the NASA administrator both attended the First National Conference on Peaceful Uses of Space in Tulsa, Oklahoma, two weeks later. They did not have a chance to talk, however. Along with Dr. Lovelace, Cobb busily answered media questions about the ongoing women's testing. She even took the opportunity to pose next to a full-scale model of the Mercury capsule on display at the conference. As the keynote speaker at the banquet, Webb spoke about Shepard's flight and the new challenges facing the space agency. And then, at the end of his prepared remarks, Webb addressed Cobb's concerns. In front of the entire audience—and without mentioning anything to her beforehand—Webb appointed Cobb an official NASA consultant. The announcement surprised Cobb utterly.[12]

There was no chance to discuss the assignment before Cobb left for Paris, and the space agency's first explanation of the appointment appeared in the press. When the agency released the formal announcement of Cobb's position, however, Webb had still not decided on the precise responsibilities she would undertake. He simply explained that she would report to Washington, DC, as a consultant. When reporters pressed the NASA administrator to specify Cobb's new duties, Webb sidestepped the issue, praising her abilities generally. According to the NASA administrator, Cobb would be "a great asset to any part of the program."[13]

Appointing Cobb a NASA consultant seemed like a good way to placate her, thereby managing the issue of women astronauts. With the new arrangement, Webb could answer any questions about female astronauts by pointing to Cobb's unpaid position, indicating that the space agency continued to investigate the issue. Moreover, offering an administrative appointment was a tried and true method of co-opting a vocal critic. If Cobb had an official affiliation with the space agency, she might be less likely to criticize it. Webb could downplay Cobb's goals without having to make a public stand against them.

Probably much to Webb's chagrin, Cobb took the appointment as an endorsement of her ideas. Indeed, she assumed that her primary task as a NASA consultant would be to serve as a liaison between the space agency and Lovelace's Woman in Space Program. Writing from Paris, she pledged to do her best to fulfill that role: "It is my heart's desire and serious purpose to see the first woman in space be an American. I assume this consultant position will be concerned with this primary objective." Cobb had very specific ideas about how to use the arrangement to help the space agency.[14]

After meeting with Webb and two other NASA officials, Cobb suggested that she could contribute to NASA on three projects: putting a woman into space before the Soviet Union; gathering medical findings on women's capabilities for space; and making the public aware of NASA's missions. Specifically, she suggested that NASA include a woman on a special suborbital mission using a Mercury Redstone rocket. To justify such a launch, Cobb cited the same reasoning that had often been used to justify women's role in aviation decades before. "This would also prove the safety, dependability and simplicity of our spacecraft systems. If a woman can do it—it *must* be safe and simple!" Naturally, Cobb volunteered herself to fly the mission. To support her candidacy, she included a one-page biography highlighting her aviation credentials.[15]

Attached to her biographical summary, on a separate page, Cobb included this prayer:

> To each is given the ability to serve. My life's purpose is to find that way in which I can best serve. God has given me the ability to fly—the unrelenting desire to traverse his skies—and the Faith to believe.
>
> I offer myself—No less can I do.

Outspoken about her faith, Cobb believed that volunteering for space was the right opportunity for her and an ordained service to her country. Indeed, Cobb's piloting ambition and her fervent religious faith explained her willingness to offer herself for a dangerous mission such as flying a suborbital space flight on a modified ballistic missile. Within the space agency, however, Cobb's vocal religiousness did not fit what NASA expected from its public representatives. Her fervor highlighted the contrast between her conviction and NASA's scientific rationality.[16]

From the very beginning, Cobb's NASA consultant position did not achieve what Webb hoped. Invigorated by the space agency's endorsement, Cobb continued her efforts to renew the women's testing and promote women as astronauts. The response she inspired pressured NASA to justify its policies. During the summer of 1961, NASA's public relations department received enough outside inquiries that the space agency's own employees also asked why women could not fly in space.

Facing several letters on his desk from citizens requesting clarification about the space agency's policy regarding women astronauts, a NASA official inquired up the chain of command about how to reply. He needed a better explanation. As he outlined the problem: "The medical portion of a selection and training program is but a guide in selecting physically and mentally qualified people, and as you know, women fall within the physically qualified; therefore, there must be other valid reasons why or why not we are to use women in our space flight program." The space agency needed to come up with a more plausible rationale for excluding women from its astronaut corps. And from NASA's perspective the public relations problem stood to get worse before it got better.[17]

"Arrangements . . . Canceled by Navy"

With the women's Pensacola examinations pending at the end of the summer, the space agency faced another series of inquiries about its all-male astronaut corps. In the end, the navy's withdrawal of its cooperation preempted the controversy that would have been raised by a dozen women's taking advanced aerospace tests. The military service did not act without consultation, however. NASA played a role in the navy's decision to withdraw its cooperation from Lovelace's Woman in Space Program.

Lovelace's informal agreements became known by navy command in early August 1961 when Jacqueline Cochran shared a car with Vice Admiral Robert B. Pirie, the deputy chief of naval operations (air). During their drive through Washington, DC, Cochran told Pirie about Lovelace's Woman in Space Program and its plans for the future. After she left the vehicle, she worried that she might not have explained the program clearly enough during their hurried encounter. To be sure that Pirie understood her, Cochran reiterated her description of the women's testing program in a two-page memorandum that she sent to his home.[18]

Cochran's communications did not betray any well-kept secret. Despite his keeping the precise details of the Pensacola arrangements relatively quiet, Lovelace's private project had received press attention throughout 1961. In addition, Cobb had kept NASA apprised of her progress all summer. Furthermore, Pensacola naval officials communicated regularly with navy headquarters at the Pentagon. Nonetheless, Pirie's personal knowledge about Lovelace's September plans for testing women at the Naval School of Aviation Medicine can be directly traced to his communication with Cochran.

As a top navy official charged with overseeing naval aviation, Pirie reacted to the news by writing to the space agency inquiring whether NASA had a "requirement," or official request, for the women's testing. A requirement served as the first step in the procedure by which government entities allocated federal funds. Without a specific requirement, federal money could not be spent on a project. The answer from NASA came back: no. Without an official request from NASA, the navy could not lend its material support to the women's testing.

The cancellation decision flowed directly from Pirie's inquiry to Cobb's final telegram. When NASA received Pirie's probe, Hugh Dryden, the NASA deputy administrator, had his staff confer with the navy about the answer. When navy officials learned that NASA had not requested the women's testing, officers at navy headquarters knew that the service would not be reimbursed for the use of its men and materiel. As a result, Pentagon staffers informed the Naval School of Aviation Medicine that it could not conduct the women's tests.

Because the cancellation derived from the absence of a document—the "requirement" or official request—when Cobb flew to Washington, DC, to investigate the problem she had difficulty figuring out how to fix the bureaucratic impasse. She learned easily that the navy's withdrawal of cooperation had postponed the planned tests indefinitely. But it was harder to rectify that situation. Determined not to give up, she stayed in the capital. Cobb recalled that during this trying time "I did a lot of wandering around Washington, a lot of thinking, a lot of praying about it. I knew it was important, not only for us, it was important for our country, so I continued to do whatever I could." Within a few weeks, she arranged a meeting with Edward Welch, the executive secretary of Kennedy's National Aeronautics and Space Council. None of the people she contacted could offer much help, however. The testing could be restarted only by the organization that had withdrawn its participation. And the navy would not act without an official reason to do so.[19]

During the first month that Cobb spent in Washington trying to restart the canceled Pensacola examinations, no written records existed to document the relevant communications between NASA and the navy. Both organizations had taken care not to commit their opposition to paper. At a basic level, the deliberate lack of a paper trail made it hard to trace the cancellation rationale. More significantly, however, officials at both agencies recognized that they could not be seen as dismissing women's concerns outright. Both organizations wanted to avoid controversies that could detract from their other advanced aviation and aerospace projects.

Finally, weeks after the navy withdrew its cooperation from the Lovelace initiative, NASA sent the navy a written confirmation explaining its decision. In a brief notice dated October 2, 1961, the space agency recapped the negotiations that had led to NASA's assertion some weeks earlier that it lacked a requirement. Hugh Dryden, NASA's deputy administrator, had received Admiral Pirie's initial inquiry in August. The admiral had asked whether the space agency had requirements for "physiological and psychological tests of a group of twelve women," to be conducted by the navy "to indicate generally the potential of women as future astronauts." Discussions ensued between the staffs of the navy admiral and the NASA deputy administrator. Dryden recorded that ultimately the space agency answered unequivocally: "NASA does not at this time have a requirement for such a program." He conceded, however, that it might investigate such issues "at some time in the future." In the meantime, NASA needed to keep the press and the public focused on its human spaceflight program and new lunar landing goal.[20]

Government willingness to sponsor women's astronaut testing had not changed significantly in the two years since Donald Flickinger and the Air Research and Development Command dropped its nascent "girl astronaut program" in 1959 owing to fears of adverse publicity. NASA officials did not want the public relations headache of explaining why they had not admitted any women into their astronaut corps. And the navy had no interest in promoting women as potential space travelers. By confirming NASA's lack of a requirement, both agencies found a way to make a troublesome problem go away. When Lovelace learned about the Pensacola cancellation, however, his problems had only begun.

Although Lovelace ran the Woman in Space Program as a part of his private namesake foundation, he also served NASA as the chairman of the agency's Special Advisory Committee on Life Sciences. As such, his work required that he remain on good terms with the space agency's administration. When he began working on the Woman in Space Program, Lovelace did not anticipate any conflicts with NASA. Once NASA officials became involved in the cancellation of the scheduled Pensacola examinations, however, he needed to repair his relationship with the agency and its leaders.

Lovelace had not worried that his private research would antagonize the space agency, for two reasons. First, when Eisenhower created NASA in 1958 to consolidate American space efforts, existing space programs did not immediately defer to the new government agency. The military services, most notably the air force, continued their own space efforts (including projects intended to launch human beings into space) for several years after 1958. In 1959, creating a space-related research project did not yet invade bureaucratic turf that had been staked out as exclusively NASA's. Second, like many other aerospace researchers, Lovelace cared more about experimental results than bureaucratic divisions. When Flickinger could not sponsor the program, Lovelace picked it up without any concern about NASA's reaction.

Furthermore, Lovelace's women's testing project began during the period of NASA's institutional flexibility before the 1961 lunar landing decision. Lovelace wanted to begin gathering the baseline medical information that the space agency would need to include women in its launches someday. Lovelace hoped that NASA might become interested in a female astronaut project once he had definitive test results. Donald E. Kilgore, a physician who worked with Lovelace at the Clinic, explained that "NASA had not been involved. NASA did not even know that this was taking place because this was supposedly secret. They did not want any publicity on it until the package was ready for sale." By the time the space agency learned about Lovelace's women's testing program in August 1961, however, the new lunar mission had streamlined the space agency. NASA had little tolerance for distractions. Although Lovelace had not intended to put the space agency—and himself—in an awkward position, he now needed to relieve the tension that his research project had created.[21]

When news of the Pensacola cancellation arrived in Albuquerque, however, the potential for a rift with NASA grew even greater because Lovelace was not present to make amends. When the navy called the Lovelace Foundation to notify the staff of the cancellation, Lovelace could not answer. He was caught in a Texas hurricane—unreachable. Lovelace's secretary, Jeanne Williams, got the news instead. She sent the telegrams to each of the women participating, telling them the Pensacola trip had been called off. Then she cabled Cochran to explain the full story: "Received word that arrangements for women to go to Pensacola cancelled by Navy Dr. Lovelace in storm area in Texas so unable to contact him at present time Have advised girls and will advise him as soon as possible." When Lovelace returned, he had a good bit of repair work waiting for him. As head of NASA's Life Sciences, he had stepped rather heavily, and somewhat publicly, on the space agency's toes by initiating a testing program that now posed a public relations problem for the human spaceflight program. Upon returning to Albuquerque, Lovelace wrote a letter to NASA administrator Webb. The timing of Lovelace's letter suggests that it may have provided the impetus for Dryden's written reinforcement of NASA's decision.[22]

What Lovelace hoped to do, however, was to avert any problems that might have been raised by the Pensacola conflict. In two full pages, he worked to anticipate any of Webb's possible concerns. He carefully explained what the program was—and what it was not. In particular, he noted that the women's testing, which he called a "research program," had been privately funded. Lovelace also took pains to clarify the rationale behind running these physical examinations: "The purpose of these comprehensive examinations here was not to try to put a woman into space at an early date but to collect detailed information on them and to endeavor to determine where they would best serve in the future." Looking at the American space effort over the long term, Lovelace speculated, space stations might eventually employ female scientific technicians to work on biological experiments.[23]

After Lovelace explained the testing program's purpose, he assured the NASA administrator that he had not intended to put the space agency in an awkward position. Primarily, he stressed that although he had tested women for fitness as potential astronauts, the pilots who underwent the examinations had never been promised a place in a female astronaut program: "All these women candidates were informed before coming here that there was no astronaut program for women at the present time and that maybe there would not be one for several years." Lovelace wanted the NASA administrator to understand that he had not led the women to expect anything from the space agency itself.

In the last few paragraphs of his letter, Lovelace suggested that it might benefit the space agency to resume the women's testing. He hoped that all the work that had been put into the program would not go to waste. Since he had already assembled a group of qualified candidates, it seemed inefficient not to complete the examinations. "There seems to be no reason why there should be any urgency about putting a woman into space at the moment, but having given a group of twenty women candidates the preliminary medical tests here with twelve of them having passed, I would hope to see the subsequent tests not delayed too long." As a scientist, Lovelace thought it was important that all the women undergo the examinations together to ensure equivalent conditions. If NASA would encourage the navy to permit its facilities to be used, the program could be completed as planned.

Lovelace argued that a female testing program could be continued without interfering with the men's astronaut program. Including women in the astronaut corps could not even be considered until scientists knew whether they could withstand the expected rigors of space flight. By completing the examinations without any more attention, researchers would be able to find out how women fared. Lovelace assured Webb that he was not interested in individual publicity: "Personally I would like to see the results of these tests, when made, kept confidential for the time being within the medical groups concerned until a scientific paper can be prepared and presented as was done with the men." He promised that if the examinations could continue, he would give NASA the test results "at no cost to the government." Lovelace hoped that Webb would not dismiss the women's testing; he wanted to complete it so that he could offer definitive findings about women's capabilities for space flight.[24]

Beyond offering these suggestions, however, Lovelace could not push too hard for NASA to authorize the continued testing. The U.S. space agency provided both work and prestige to the Lovelace Foundation. In addition, Lovelace's own affiliation with the agency permitted him to work on achieving piloted spaceflight. He needed to maintain a working relationship with NASA and its administration. As much as he wanted to continue the women's testing, Lovelace could not be the primary advocate for resuming the Pensacola testing because he had other space projects to maintain, including other NASA programs. Advocating for women astronauts fell to Cobb, NASA's newest consultant.

"Not *Yes* and Not No"

Despite Cobb's consulting position with the space agency, it quickly became apparent that NASA had no intention of following her recommendations. In the face of this resistance, Cobb adopted a new attitude that defined her political advocacy. Until she received a definitive denial and a sufficient explanation, she would not be discouraged. Over the next year; Cobb's strategy allowed her to continue advocating for women astronauts in the face of the mixed messages being publicized by the federal government about women's place in the U.S. space effort.

By the end of 1961, NASA administrator Webb concluded that Cobb, whom he had previously touted as "a great asset to any part of the program," did not have much to offer the space agency after all. NASA had already established its plans for fiscal year 1963, by which time the space agency would be actively preparing for the moon missions. To that end, in December Webb made final arrangements with the Department of Defense for the two federal organizations to cooperate in testing, selecting, and training the Apollo Program astronauts. The Department of Defense also planned to take charge of the "interim flight program" during the selection process. As a result, the entire astronaut program would soon be focused on producing crews for the two-astronaut Gemini project and the three-astronaut Apollo moon landings. In December 1961 these plans had already been established, although NASA and the Department of Defense had not yet announced them.

On December 15, Webb wrote to tell Cobb that he had reconsidered her consulting position. With the Defense Department agreements in place, Webb did not see her fitting into NASA's upcoming plans: "Under the circumstances, and since we have not found a productive relationship that could fit you into our program, I am wondering if there is any advantage to continue the consulting relationship that you and I had in mind." Hoping to let her down easily, Webb worded the letter indirectly, expressing regret that Cobb did not fit into the new space agency arrangements for the coming months and years: "Those in charge have not come up with any assignment that I thought we should ask you to undertake." He asked her to contact him when she had considered his assessment. Webb expected Cobb to defer to his declaration, understanding that her official relationship with the space agency had been terminated.[25]

Notably, however, Webb did not explicitly rescind the consulting appointment. Thus Cobb did not consider the arrangement to be over. In her battle with the bureaucrats who blocked women's path to space, Cobb had adopted the attitude that until she received an explicit no, she would persevere. Furthermore, Cobb refused to accept any rationale that could not be explained to her satisfaction. As a result, the polite rebuffs and explicit brush-offs penned by NASA officials did not cause her to surrender. She grew frustrated with her lack of progress, but she did not give up.

Regardless of Webb's December letter, Cobb began the new year of 1962 working actively in her role as an unpaid NASA consultant. She crisscrossed the country, speaking out for female astronauts. In February 1962 she wrote to her "fellow lady astronaut trainees," or FLATS, to update them on her progress. "The past four weeks I have attended NASA meetings, briefings, astronautics meetings, scientific symposium[s], given speeches and talked, argued, reasoned, and listened to many of the top people in our country's space programs." Despite the lack of progress, she pursued a full schedule. At the end of the month she spoke at the NASA-sponsored First Woman's Space Symposium in Los Angeles. The very existence of such a forum—sponsored by the U.S. space agency, no less—indicated that NASA had begun to include women's concerns in its public discussions of the space effort. Nonetheless, no one in the space agency would explain how women could become astronauts. As Cobb wrote to her fellow pilots, "The answers I got were more of the same and typical of bureaucrats—not yes and not no." Without a definitive denial, however, Cobb kept going.[26]

As Cobb learned the intricacies of negotiating national politics, she translated her religious practices into a political approach. She believed that as long as she maintained her faith, she could not be denied. As she wrote to her fellow lady astronaut trainees, "I am as tired and discouraged about this as I know each of you are—but I will keep fighting for it, cause I believe in it. It's an uphill battle but with faith and persistence I still believe it will be done." Resolute in her belief that her goals advanced the nation's space efforts, Cobb refused to relent.[27]

Cobb's conviction that not everyone in the federal government opposed women astronauts—and that advocates existed who might speak on women's behalf—did not arise only from blind faith. Indeed, the president and the vice president continued to invite women to lend their skills to the national space effort. Despite NASA's lack of interest in female astronauts, the Kennedy administration's public discussion of the space program encouraged both women and men to participate in America's space effort.

In December 1961, the same month when Webb suggested ending Cobb's consulting appointment because no programs needed her input, a feature page in *American Girl* magazine posed an invitation from President Kennedy to girls interested in space. Under the headline "Lots of Room in Space for Women," the appeal appeared superimposed on the photo of a rocket rising from a launch pad. The "Message to You from the President" reminded women and girls of the "new and critical importance" of the space effort. Kennedy urged girls to think about participating in America's space program because "the skills and imagination of our young men and women are not only welcome but urgently sought in this vital area." Just three months later, Vice President Johnson called for high school students to study math and to consider pursuing engineering in order to fill "the nation's urgent need for college-trained men and women in the field of engineering." Whether or not the space agency welcomed women into the astronaut corps, the Kennedy administration invited them to participate in the space effort.[28]

Judging from letters written to the vice president, other women besides Cobb took this rhetoric seriously. In March 1962 Catherine Smith, a married African American woman from Pontiac, Michigan, wrote to Johnson volunteering to be "one of the First American Women in Space Program." Within a week, the vice president's office also received a letter from thirteen-year-old Fritzi Mann, asking what kinds of classes she should take so that she could go into space someday. Mann may have been inspired by Johnson's call for girls to study engineering, but she clearly envisioned herself as an astronaut, not sitting at a desk calculating trajectories.[29]

Despite their appeals to women and girls, neither Kennedy nor Johnson intended to encourage the idea of female astronauts. Rather, they hoped to employ women as support staff in space-related industries or, if they possessed mathematics or engineering degrees, as "human computers." They also wanted to make sure that female voters felt connected enough to the space effort to support massive allocations of taxpayer funds. Nonetheless, the Kennedy administration's public appeals contradicted NASA's actual policy. To decode these mixed messages, Cobb needed someone who knew her way around Washington politics.

Fortunately, one of the women slated to join Cobb in Pensacola was Jane B. Hart, the wife of Democratic senator Philip Hart of Michigan. In March 1962 Hart joined Cobb in working publicly for the resumption of the Pensacola tests. Cobb enthusiastically conveyed the news to the rest of the Lovelace candidates: "Mrs. Janey Hart, wife of a Michigan Senator, and mother of eight children has shed her anonymity as one of the women pilots to pass the Lovelace test in order to help our campaign to get a program started for 'lady astronauts.'" Hart began writing letters to each member of the Senate and House space committees. With her letters, she included copies of Cobb's February speech at the First Women's Space Symposium. Using her knowledge of the nation's capital as well as political connections she had fostered during her participation in her husband's career, Hart began to lobby government officials to resume women's astronaut testing. In 1961 and 1962, Cobb and Hart kept the question of women's capabilities as potential astronauts on the national agenda despite the considerable efforts of others in the federal government to make it go away.[30]

Through her husband's Senate office, Hart requested a meeting with Vice President Johnson. As a fellow Democrat and a political friend to women, Johnson seemed a likely ally. In his position as the head of the National Aeronautics and Space Council, he had the clout to help them. Hart did not ask for a full-fledged women's astronaut program. She just wanted the vice president to help the Lovelace group complete the Pensacola tests. She recalled, "I was hoping that he could get us through the next level. Because it seemed to me that if we got through that, that we would be pretty convincing." In retrospect, however, she described her appointment with Johnson as "a miserable meeting, actually. Because he was terribly uncomfortable. . . . He kept saying there wasn't much that he could do about it."[31]

This was not true. Johnson might have offered some assistance; he chose not to. Although he acted noncommittal when he met with Cobb and Hart, actually he vehemently opposed their cause.

7

"A FACT OF OUR SOCIAL ORDER"

Jerrie Cobb, John Glenn, and the House Subcommittee Hearings

Before his meeting with Jerrie Cobb and Jane B. Hart, Lyndon Johnson's assistant, Liz Carpenter, offered the vice president an easy way to make a gesture on the women's behalf. As she assembled newspaper clippings and other briefing materials, she also drafted a letter from Johnson to the NASA administrator. The note did not endorse female astronauts but instead offered a carefully worded inquiry asking "whether NASA has disqualified anyone because of being a woman." Carpenter recommended that Johnson sign the letter during the upcoming appointment. In that way, she suggested, he could show Cobb and Hart a visible sign of his cooperation.[1]

In her cover memorandum outlining the women's cause, Carpenter encouraged Johnson to use the letter, arguing, "I think you could get a good press out of this if you can tell Mrs. Hart and Miss Cobb something affirmative. The story about women astronauts is getting a big play and I hate for them to come here and not go away with some encouragement." After the meeting, Johnson could mail the letter to the NASA administrator knowing that it did not demand real action but only requested clarification.[2]

When Carpenter composed the letter for Johnson, however, she unknowingly created the only piece of written evidence that testified to how strongly key government officials opposed women in space. Johnson did not sign Carpenter's draft. Instead, he added a different message in the signature line. In uncharacteristically emphatic inch-high handwriting, he scrawled across the width of the page: "Let's stop this now!" He then marked the sheet to be filed without further action. Vice President Lyndon Johnson objected to the woman in space initiative.[3]

Except for Johnson's handwritten outburst, government officials left no other written evidence of private opinions about women astronauts. The previous autumn, navy and NASA officials had pointedly avoided recording their involvement in the Pensacola cancellation. And although NASA officials actively discouraged Cobb's overtures, they never committed the space agency's assessment of her efforts to paper. Johnson's exclamation, buried in his files for almost forty years, betrayed how vehemently he opposed the cause of women in space.

Political support for female astronaut testing would not come from the executive branch. Instead, in the year after a bureaucratic impasse canceled Lovelace's Pensacola testing, the little help that Cobb found in Washington, DC, came from the House of Representatives. In 1962 the House Committee on Science and Astronautics convened a special subcommittee to investigate women's place in the American space effort. Like Johnson's private exclamation, the records of the 1962 House subcommittee hearings reveal the private motivations of the historical actors who argued for and against resuming women's astronaut testing. In doing so they offer new insight into the cultural politics surrounding women in space in the early 1960s.

Cordial "Dear Jerrie" Letters

Despite the vice president's public reluctance and private opposition, the issue of possible female astronauts stayed on his desk, thanks to continued efforts by Cobb and Hart. Within a week of her meeting with Johnson, Hart addressed five hundred women at the New Century Club in Wilmington, Delaware. Her speech describing her success in the Lovelace tests prompted one of her listeners, Mrs. George B. Ward Jr., to write to the vice president urging him not to neglect women's potential: "We want to do something more important than drink tea, play bridge, and sit on the side lines while there are vital things to do."[4]

Ward and others who wrote to Johnson about women astronauts each received a polite reply explaining that "there are no standards for selection that exclude women, but the requirements are so exacting that only trained and experienced high-performance jet aircraft test pilots can be expected to succeed." Letter writing would not change the vice president's adamant opinion; the two women pilots would have to find other allies. They lost no time. In the months after her meeting with Johnson, Cobb engaged in a flurry of meetings and correspondence with influential figures who might share her interests.[5]

In her eagerness to create a groundswell of support for her cause, however, Cobb began to report that she had found allies where she had none. Whether intentionally or not, Cobb often misrepresented the opinions of key decision makers in her letters. In retrospect, it remains difficult to distinguish between Cobb's own understanding of these interactions and her interpretation of them as a part of her strategy of never taking no for an answer. Did she embellish her assessments as a strategy to gather allies to a winning team or, in her earnest desire to find advocates, did she hear genuine support in political pleasantries? Whatever her motivations, during spring 1962 Cobb began repeating her overly rosy assessments of political support to other key figures.

Unaware of the vice president's handwritten outburst and less sensitive to Johnson's uncomfortable demeanor than Hart was, Cobb interpreted his lack of overt opposition during their March appointment as a sign of encouragement. In a letter reporting her efforts to the other Lovelace candidates, she described their meeting as a success: "Janey had arranged for us to have a conference with Vice President Johnson, who was very receptive to the idea and said he would check into the matter. . . . However, because of his in-between position on space matters (NASA, Space Council, President) he was emphatic about not being quoted." Cobb went so far as to repeat her assessment of Johnson's support to NASA representatives.[6]

In a letter to Wernher von Braun, the charismatic architect of the American rocket program, Cobb confided, "I've talked to many people in Washington including Vice President Johnson who thinks it's important that the United States put the First woman in space, but unless we start soon, Russia's going to beat us again." Cobb's claim seemed so out of sync with the general sentiments being expressed within government circles, however, that the public information officer who initially reviewed Cobb's letter seemed confused by her claims of support. When he presented the note to von Braun, he inquired cautiously: "In the letter she quoted Dr. Stuhlinger and The Vice President as approving. Before I prepare an answer I just wanted to know if you wanted to go on record as approving lady astronauts or whether the letter should be non-committal." Recognizing the delicacy of the political situation, von Braun opted to reply in a gently worded rebuff.[7]

The rocket scientist carefully avoided stating whether he personally approved of women as astronauts. He instructed the aide, "Suggest we make the answer in the form of a cordial 'dear Jerrie' letter." To soften the blow, von Braun expressed his sympathy with Cobb's desire to fly in space because, as he wrote, "I frequently have the feeling that I would like to make the trip too." Although von Braun reiterated NASA's position that "jet-test experience" remained a critical prerequisite to spaceflight, he assured Cobb that he felt she would probably make a fine astronaut.[8]

In Cobb's optimistic readings, she received significant approval for women in space during this period but nonetheless made frustratingly little progress with NASA. She began addressing her appeals to other space agency officials besides NASA administrator James Webb, hoping for a different answer. Although the responses continued to cite the necessity of NASA's jet test piloting requirement, as Cobb polled other space agency representatives, some replies revealed additional reasons why the space agency did not pursue women astronauts in the early 1960s. As Robert Gilruth, director of NASA's Space Task Group overseeing Project Mercury, stated straightforwardly, "We cannot select women just because they are women." Even if the Soviets achieved the feat of flying a woman in space first, he argued, the United States was competing for "sound technical and scientific information," not "propaganda stunts." He reasoned, "At such a time as the ability of women to operate efficiently in space becomes an important enough question, we will undoubtedly investigate the problems involved."[9]

Despite the repeated inquiries that Cobb sent to space policy makers, none of them thought her cause was "important enough" to treat as anything other than as a public relations problem. Without an organized movement calling for women's full participation in public life, women's issues lacked clout. As a result, although space agency officials and other space policy makers had to tread carefully in responding to Cobb's repeated inquiries, they did not have to change their answers. If Cobb wanted a different response, she needed someone who could persuade the space agency to alter its astronaut qualifications. Or she needed an ally who could persuade Pentagon officials to allow women to use military aerospace testing facilities without a NASA requirement. For more than six months after the Pensacola cancellation, neither seemed possible.

A key meeting with a House committee chairman finally broke the logjam. Cobb met Rep. George Miller of California at the Goddard Memorial Banquet. Unlike other politicians who rebuffed her, Miller took her seriously. More important, as the chair of the House Committee on Science and Astronautics (known popularly as the House Space Committee), he was in a position to help. He offered to investigate the issue.

Despite many assumptions that Jane Hart's congressional connections must have been the impetus for the 1962 House subcommittee hearings, in fact Cobb's meetings with key representatives on the House Space Committee actually prompted the formation of the special subcommittee. After agreeing to investigate Cobb's concerns, Miller turned responsibility for the issue over to Republican congressman Victor Anfuso of New York. Miller, the full committee's chairman, had agreed to convene the subcommittee hearings, but he would not attend them. Instead, his colleague tracked Cobb down while she was in New York City. At Anfuso's invitation, she flew back to Washington to have lunch with him and an official from the State Department's foreign affairs department. She also visited one of the committee's hearings. In June, the House Space Committee announced that a special subcommittee headed by Anfuso had been formed; public hearings about astronaut qualifications would be convened in July.[10]

"A Properly Organized Astronaut Program for Women"

In the weeks between the official announcement of the hearings and their opening date, private negotiations began behind the scenes as the interested parties prepared their separate testimonies. Hart recalled that Cochran contacted her to discuss the hearings. "I got a call from Jacqueline Cochran to come to New York. She wanted to talk to me about it. So I went up and had lunch with Jacqueline and her husband at her apartment. And it was clear to me that she was not going to support it." Although she and her millionaire husband had funded Lovelace's testing initiative fully, Cochran did not support resuming the Pensacola examinations. Her contested role in Lovelace's Woman in Space Program had left her with different ideas about when and how women should be added to the U.S. space program.[11]

Cochran's concerns about the female astronaut testing program—and her place in it—had festered for some time. When the conflict about her peripheral role came to a head in May 1961, her husband Floyd Odlum had interceded, implicitly threatening to pull the funding for the program. Although Lovelace made a concerted effort to include Cochran in the program from that point on, her interest never fully recovered. The Cochran-Odlum Foundation still defrayed each woman's expenses in Albuquerque and supplied funds to cover the Lovelace Foundation's costs. But as the Pensacola tests approached, Cochran could see the program's direction changing. If the Lovelace women became recognized as potential astronaut candidates, her role would diminish significantly. In addition, more attention would shift to the leader within the group: Jerrie Cobb.[12]

In the months before the scheduled Pensacola tests, during the summer of 1961, Cochran had attempted to reassert her leadership in Lovelace's testing. In a letter to Lovelace, she complained that "it is apparent that one of the girls has an 'in' and expects to lead the pack." Given their long friendship and previous history regarding the Woman in Space Program's leadership, Lovelace would have recognized the implications of her statement. Just in case he did not, Cochran made her opinion clear: "Favoritism would make the project smell to high heaven." Jerrie Cobb's public recognition as the leader of the group that Cochran had funded—and her recent appointment as a NASA consultant—was galling.[13]

Several weeks later, Cochran again picked up on the theme of Cobb's publicity, this time sharing her opinions with the Woman in Space Program participants. In July 1961, a week after the Pensacola tests were postponed for the first time, she sent a long letter to each of the women explaining her participation in the program and her hopes for its future. Cochran cautioned against anyone's misconstruing the new phase of examinations as a promise of a forthcoming organized program. If something did develop, however, Cochran volunteered to lend her experience to leading it. She explained, "I think a properly organized astronaut program for women would be a fine thing. . . . It is possible that my previous experience with group efforts by women in the air can be of some value." Cochran cautioned each of the women to avoid individual publicity regarding the tests.[14]

By the next spring, her feelings had not changed. This time, however, she told Cobb directly. During the delays as they waited to witness John Glenn's February launch, Cobb dined with Cochran at Cocoa Beach. Cobb brought along Jane Rieker, a *Life* magazine writer who was working on an "as told to" autobiography of Cobb (another sign of upcoming attention to the young pilot). During the course of the evening, Cochran grew frustrated because Cobb did not seem willing to hear her opinions regarding women astronauts. Finally, on March 23, 1962, Cochran decided to clarify her views. She composed a dense four-page letter explaining to Cobb that women's space travel would probably have to wait until the men's program was well established. As

Cochran wrote, "Their [women's] time will come and pushing too hard just now could possibly retard rather than speed that date." Cochran thought that Cobb's personal publicity damaged Lovelace's program. She closed the letter by advising, "If you go along with the soundness of the group idea, as I hope you do, then you can be particularly helpful to a program by going out of your way to create the group image publicly." Cobb's personal campaign to be the first woman in space made Cochran uncomfortable.[15]

As a result, given her persistent dissatisfaction with the public image of Lovelace's Woman in Space Program, Cochran saw the House subcommittee hearings as a chance to resurrect the women's astronaut testing program in a form more to her own liking and one that would be, in her opinion, more likely to succeed. Before the 1962 hearings convened, Cochran assured key space policy makers that she intended to testify that no program should be pursued that might interfere with NASA's current missions. In the month before the hearings, Cochran let influential decision makers know about her letter to Cobb. She distributed copies to Vice President Johnson, Admiral Pirie, General Le May, Dr. Lovelace, and Rep. Olin Teague. NASA deputy administrator Hugh Dryden and Robert Gilruth, director of NASA's Manned Spacecraft Center, also received copies of the text. Each time, she made a point of distinguishing her views from Cobb's.

Consequently, even before the congressional subcommittee hearings convened, NASA insiders and military officials already knew about Cochran's moderate stance on women astronauts. She reinforced this alliance by distributing copies of her prepared testimony through the space agency for review in the weeks before the hearings. She sent individual copies to Webb, Dryden, and Gilruth at NASA. Her old friend Gen. Curtis Le May, the air force chief of staff, also previewed her statement.[16]

The recommendations Cochran developed in her testimony included plans for a large, comprehensive women's testing program. She wanted to lead women pilots into space in the same way that she had marshaled female fliers during World War II: through a widespread, organized effort. Although Cochran would testify that the Pensacola testing should not be resumed if it would interfere with NASA's current programs, she did not intend to give up on women as astronauts. Rather, she hoped that a strategic alliance with the space agency would allow her to complete the testing that Lovelace had begun. This time, however, the project would take a different form and carry NASA's endorsement, if not the agency's outright sponsorship.

When Lovelace previewed Cochran's prepared statement, he sent her two pages of suggestions, endorsing her ideas for an expanded program by stressing how long any significant research project would have to be. He concluded, "To have meaningful data that we can prove out would probably take five years." As someone deeply invested in NASA's human spaceflight programs, he agreed with Cochran's willingness to wait: "We do not recommend injecting women into the present program." Because Lovelace would not participate in the subcommittee process, his endorsement encouraged Cochran. Indeed, her strategy of circulating her testimony ahead of time uncovered a lot of support. Each of the various officials who previewed Cochran's subcommittee testimony applauded her argument that women should not be added to the space program if it risked detracting from NASA's main objectives. From the perspective of space policy makers, having the estimable Miss Cochran offering such testimony at the hearings made their case easier.[17]

Within the space agency, NASA officials prepared for the hearings by scrambling to get up to speed on a controversy that had been deemed irrelevant until the public hearings loomed. With Cobb's letters continuing to come in and Cochran's opinions also being circulated, the conflicting correspondence regarding women astronauts prompted the director of manned space flight to attempt to sort them out. He compiled the letters into a packet with a two-page memorandum briefing the associate administrator on the ongoing debate. In these briefing materials, Cochran's letter to Cobb became a key piece of evidence used to counter Cobb's public appeal for a crash program to put a woman in space first.[18]

In the meantime, the NASA administrator maintained the space agency's public relations campaign, trying to keep the focus on existing programs. In a speech before the General Federation of Women's Clubs in Washington, DC, on June 27, 1962, Webb stressed that because of the limited number of space flights planned, the astronaut's role in each flight remained vital. Consequently, he argued, NASA required stringent astronaut qualifications, requirements that were "more readily met by men than by women."[19]

While Webb tried to persuade women that astronauts needed to be men, Cobb and Hart prepared to convince congressional representatives that women could fill the astronaut's role just as well. With only short notice to prepare their speaking points, however, the women found it difficult to plan. The subcommittee's schedule called for testimony from Cobb, Hart, and Cochran, followed by statements from the NASA representatives. Hart received notice of the subcommittee hearings only days before the starting date. As a result, she and Cobb prepared their statements separately, with just one day to coordinate their testimony before the hearings began. They hoped they would be able to impress the representatives with the strength of the testing that had been done by the Lovelace Foundation. For them, the best possible result from the hearings would have been the resumption of the Pensacola testing.[20]

Before the House subcommittee hearings ever opened, however, a number of significant decisions had already been made that hindered their cause. Because of his other affiliations with the space agency, Dr. Lovelace remained conspicuously absent. His appeal to Webb had been his last attempt to restart the women's testing. The navy could not permit the Lovelace group to use its facilities without an official requirement. And the agency most likely to require such testing still did not consider women astronauts an "important enough question" to investigate. Webb's decision to release Cobb from her appointment as a NASA consultant reaffirmed the space agency's determination to avoid the female astronaut question. Finally, potential advocates such as Vice President Johnson vehemently opposed the cause. If the Pensacola tests were to be reinstated, Cobb and Hart needed to find a way to overcome—or circumvent—these obstacles. A strong showing before the House subcommittee might earn them political allies and sway public opinion. It could only help.

The House Subcommittee Hearings

On Tuesday, July 17, 1962, Rep. Victor Anfuso of New York called to order the special subcommittee of the House Committee on Science and Astronautics. Significantly, the hearings investigated sex discrimination two full years before it became illegal. (Only after a southern legislator inserted the word "sex" in Title VII of the 1964 Civil Rights Act did women gain legal protection against unequal treatment based on sex.)[21] As a result, all the participants faced difficult tasks. NASA officials needed to defend the agency against discrimination charges when the agency had not broken any law, or even had direct dealings with Lovelace's Woman in Space Program. Indeed, the space agency's role in the Pensacola cancellation resulted from the absence of a requirement, not from any positive action. On the opposite side, Cobb and Hart needed a congressional subcommittee to compel NASA to change its policies without any law to back their claims. Furthermore, they wanted the representatives to pressure NASA into restoring the lost Pensacola opportunity when space agency officials could argue—quite accurately—that they had not caused the cancellation.

Cobb and Hart opened the first day by testifying that women brought valuable talents to the national space effort. In their prepared statements, they briefed the subcommittee on the history of Lovelace's Woman in Space Program. To emphasize the impressive credentials of the women involved, they submitted biographies for the *Congressional Record* of each of the women who had passed the Lovelace tests. As Hart stated, "I am not arguing that women be admitted to space merely so they won't feel discriminated against. I am arguing that they be admitted because they have a real contribution to make." Following the opening statements, Cobb showed the committee a slide program illustrating the tests the women had undergone, demonstrating that the female pilots had been able to withstand anything the aerospace physicians threw at them.[22]

In their arguments, both women relied on analogies with women in other frontier explorations in American history. As Cobb argued in her opening statement, "There were women on the *Mayflower* and on the first wagon trains west, working alongside the men to forge new trails to new vistas. We ask that opportunity in the pioneering of space." They saw space as another frontier that women could help to open. Although the prepared statements went well, when the question-and-answer period began their testimony became more complicated.[23]

In one of the first exchanges, Cobb seemed fearful of speaking beyond her own expertise. Rep. James Fulton of Pennsylvania repeatedly tried to get her to testify about women's equal competence as pilots. Cognizant of how often her interviews had been misconstrued in the press, Cobb hesitated to say anything she could not demonstrate personally. Fulton prodded her to state her knowledge in a way that could be used to strengthen her cause: "I am trying to support you. Women are competent to operate aircraft and as far as you know there is no greater incidence of accidents among women than men is there?" But Cobb would not answer in a declarative statement, even after Fulton encouraged her to state her agreement for the record: "I am trying to get you to say it." She did not want to undercut their case by making a mistake, protesting, "I don't have anything to back it up with." Finally, Hart stepped in to answer the question with information from the National Safety Council. Without other experts scheduled to testify, however, Cobb and Hart had to make their entire arguments themselves.

For Cobb, that was a daunting task. Throughout her public campaign for women in space, Cobb forced herself to overcome tremendous shyness. In fact, in the midst of the public hearings, one of the congressmen asked her how she could be brave enough to go into space when he had overheard her saying before the meeting that she was "scared to death" of appearing before the panel. Cobb deftly turned the pointed question to her advantage, answering to appreciative laughter that "going into space couldn't be near[ly] as frightening as sitting here." Although she appeared hesitant at some points in her testimony, in the end the real issue would be not style but substance.[24]

Throughout the first day of hearings, Cobb and Hart wrestled with the complicated business of acting as their own advocates. In questioning the necessity of the requirement for an engineering degree or jet test piloting experience, Cobb and Hart challenged NASA's judgment in selecting its astronauts. Moreover, in describing the cancellation of the Pensacola tests because of the lack of NASA's requirement, Cobb blamed the space agency for preventing the navy from working with her and the other women. At that point Chairman Anfuso chided her, "Will you permit me to say this to you, Miss Cobb and Mrs. Hart, that this committee has the assurances that NASA wishes to cooperate and is cooperating. . . . And I know that you don't criticize any branch of the Government. You just want answers." Chastened, Cobb replied, "I certainly do not want to criticize. I would like very much to work with NASA." But throughout the morning Cobb and Hart did find themselves in the awkward position of criticizing the agency whose cooperation they needed.[25]

Moreover, Cobb and Hart struggled to make a clear case for why, in a political climate that did not value women's participation, women should be astronauts. They could make a strong argument based on the Lovelace program's results that women possessed the physical requirements to be considered as potential astronaut candidates. Since NASA already had qualified astronauts, however, the agency felt no need to change its requirements to include female candidates. When the representatives prompted Cobb and Hart to argue that in fact women could be superior astronaut candidates, both demurred.

Neither woman wanted to go on record testifying that women made better astronaut candidates than men did. Both knew that such an argument would only be used against them later. Furthermore, they had no way of proving such claims, especially when it would require demonstrating women's superiority to national heroes such as NASA's Alan Shepard and John Glenn. As a result, although Cobb and Hart argued persuasively that women could be astronauts, they struggled to explain why women *should* be astronauts.

The presence of two women on the eleven-person questioning panel did not seem to help their cause. Rep. Corinne B. Riley of South Carolina and Rep. Jessica Weis of New York probably drew their subcommittee assignments because they were the only women on the full twenty-eight-person committee. On the subcommittee, however, the most vocal support for Cobb and Hart came from Rep. James G. Fulton of Pennsylvania. Indeed, his advocacy for the women's cause prompted teasing from his colleagues that his bachelor status made him particularly sympathetic to women's concerns.

When each of the representatives had questioned Cobb and Hart, Chairman Anfuso called a brief recess to introduce himself to Jacqueline Cochran, who had made her entrance halfway through the representatives' questioning of Cobb and Hart, interrupting the testimony. When the hearings reconvened, Cochran's prepared statement said exactly what those who had previewed her testimony expected. She argued that proof of women's physiological and psychological suitability "should not be searched for by injecting women into the middle of an important and expensive astronaut program." Instead, Cochran advised that a large experimental program would be needed to determine whether women could eventually be trained for spaceflight. She supported her claims with examples from her own considerable aviation background.[26]

Ironically, Cochran used her World War II experience with the Women Airforce Service Pilots to argue against resuming the Lovelace program. Rather than using the WASP's tremendous success to argue for women's demonstrated competence in the physical and mental demands of wartime aviation, Cochran spoke about the limitations she had encountered. Specifically, she testified, "You are going to have to, of necessity, waste a great deal of money when you take a large group of women in, because you lose them through marriage." According to Cochran, the attrition rate among the WASP owing to marriage had been "somewhere in the neighborhood of 40 percent." Regarding the Lovelace experiment, she worried that with such a small initial group, valuable funds would be wasted if even one woman chose to leave.[27]

She also suggested that rushing to have the first woman astronaut could set the United States up for embarrassment. "I would rather see us program intelligently and with assurance, and with surety, than to rush into something because we want to get there first." Throughout her testimony, Cochran contrasted her plans for a comprehensive, long-term women's testing program involving military cooperation with the resumption of the Lovelace program, a move she repeatedly characterized as a "crash" program focusing on too small a group. Such a project, she suggested, would interfere with NASA's space efforts.

By noon, the representatives had questioned all three women thoroughly. When the hearings adjourned for the day, a partial judgment of how well Cobb and Hart had used the opportunity to publicize their cause came in the press coverage of the day's events. Overall, they achieved mixed success. Newspaper stories reported both Cobb's and Hart's testimony about why women should be included in the space effort and Cochran's arguments about women's washing out of expensive training.

The pictures accompanying stories across the country also carried mixed messages. Although the photograph sent out over the Associated Press wire showed Cobb, Hart, and Anfuso posing with a model rocket, United Press International chose to illustrate Cobb's and Hart's time before the House subcommittee in a way that undermined their authority. The UPI photographer caught Cobb as she slipped her feet out of her high-heeled shoes under the table. Similar shots, which distracted from the substance of a woman's contributions by focusing on a momentary slip of etiquette, had been used for years to make military women and other women who challenged traditional men's realms look ineffective. The shot undercut Cobb's message about women's competence.[28]

Although some of the reports of their testimony eroded the power of their message, Cobb and Hart had gotten their arguments heard by national decision makers and reported in national media. Over the course of the first day's testimony, some of the representatives clearly sympathized with their cause. On the second day of the House subcommittee hearings, however, NASA brought in brought in the agency's biggest stars to testify against continuing the women's testing.

"It May be Undesirable"

The NASA spokesmen who testified on the second day of public hearings demonstrated how seriously the space agency took the House subcommittee. Fresh from the two most recent Project Mercury flights, astronauts John Glenn and M. Scott Carpenter represented the space agency. Along with George Low, NASA's director of spacecraft and flight missions, the three NASA representatives testified that women could not meet NASA's astronaut qualifications. Furthermore, they argued, beginning a female astronaut testing program would hinder other NASA projects. NASA wanted to end this distracting debate.

Low, accompanied by astronauts Glenn and Carpenter, repeated the space agency's determination that all astronauts must be jet test pilots first. Astronauts had to be able to make snap judgments in hazardous situations, a skill learned quickly in test flying or combat service. Astronauts also had to be able to take over manual controls, flying the spacecraft in case of a system malfunction. Finally, the men argued, astronauts needed the psychological stability to handle such situations calmly. Through their survival in a competitive and dangerous profession, jet test pilots had already demonstrated these characteristics.

Notably, the space agency refused to take any responsibility for establishing requirements that, although they did not obviously discriminate against women, could not be fulfilled by any American women at the time. In 1962, no women had graduated from military jet test pilot schools. After the WASP disbanded in 1944, the military barred women from flying its aircraft. Indeed, the only woman who could be categorized as a jet test pilot at this point was Jacqueline Cochran, who had broken the sound barrier in 1953, when her husband's financial interest in the aircraft's manufacturer opened the opportunity for her. The women Lovelace tested—experienced pilots with commercial ratings—could not possibly have the experience NASA required. The best that women pilots could offer in 1962 was thousands of hours of flight time in propeller-driven aircraft.

In the second day's testimony, astronaut Scott Carpenter testified that the women's extensive private aviation experience could not prepare them adequately for space flight. He argued that what a pilot learned through jet test piloting could not possibly be acquired through any other type of aviation. As Carpenter told the subcommittee, "A person can't enter a backstroke swimming race and by swimming twice the distance in a crawl, qualify as a backstroker. I believe that there is the same difficulty in the type of aviation experience that 35,000 hours provides a civilian pilot and the experience a military test pilot receives." In the hierarchy of aviation technical knowledge, jet test piloting occupied a level that was inaccessible through any other type of flying.[29]

NASA wanted jet test pilots, at least in part, because the status associated with their rank enhanced the astronauts' image. Cobb and Wally Funk, another successful Woman in Space Program candidate, discussed pooling their money to buy a single-place British jet until they learned that NASA would count only military jet test pilot experience. Flying a privately obtained jet would not fulfill NASA's jet piloting requirement. For NASA, having its public face associated with aviation's elite reinforced its image as the premier aerospace agency. Conceding that women might be able to perform the same tasks using extensive nonjet flying experience would undercut that status.[30]

Even investigating whether nonjet test pilots could carry out astronaut tasks would hamper current U.S. plans, according to the NASA representatives. Deciding to put women through aerospace fitness testing at military bases would interfere with NASA's ongoing programs because those facilities were already overtaxed by the demands of ongoing projects. The space agency's new class of astronauts also needed access to the facilities. A completely separate women's testing program would still affect the American space effort.[31]

The NASA representative assured the subcommittee members that the space agency had not intended to discriminate against women. Any woman who fit NASA's stated criteria would be welcomed. But the agency's standards needed to be maintained. As Glenn testified, "If I am on a space mission and I have someone with me, I want the highest qualified person I can [get] over there, whether it is a woman—without regard to color, race, creed or whatever it is." In fact, Glenn argued, "I am surprised that . . . instead of trying to reduce our qualifications to a lower level . . . perhaps we should be upping the qualifications and saying that we have to have test pilots with doctorate degrees and with even more experience." With multiple-person flights on the schedule for Project Gemini and the Apollo lunar landing program, the astronauts argued, the space agency needed to maintain the exclusivity of the astronaut corps. Besides, they continued, passing the Lovelace physical exam did not qualify anyone for further consideration as an astronaut candidate. Those tests simply identified anyone whose physical defects or weaknesses would disqualify them. Glenn made the case by joking, "My mother could probably pass the physical exam that they give preseason for the [Washington] Redskins, but I doubt if she could play too many games for them."[32]

Throughout the two days of subcommittee hearings, similarly patronizing comments showed that women were not taken seriously. Even representatives who supported the women's cause suggested that good reasons to include women astronauts on a flight crew might be because they are good company or because they might defuse tensions between men, who apparently would be ashamed to fight in front of a woman.

Despite the jokes, one critique that might have been expected was not used. Throughout their testimony, the NASA representatives never argued that women were too weak to be astronaut candidates. Nor did they say that women lacked the mental stability or intellectual capacity for spaceflight. Low and the astronauts recognized that dismissing women in this way would be politically unwelcome. The women's liberation movement remained two years or more in the future, but the shifts in gender relations that would make it possible were already evident.

In perhaps the most telling commentary during the two days of hearings, Glenn testified, "I think this gets back to the way our social order is organized really. It is just a fact. The men go off and fight the wars and fly the airplanes and come back and help design and build and test them. The fact that women are not in this field is a fact of our social order. It may be undesirable." Despite the absolute nature of his description—"it is just a fact"—Glenn felt compelled to explain, and thus to reassert, men's and women's appropriate roles. Glenn's testimony did not describe the existing social order. His statements ignored the history of the Women Airforce Service Pilots and the consequences that disbandment had for women pilots. Indeed, his assessment omitted women's aviation altogether. Rather than simply describing rigid postwar gender roles, Glenn was attempting to reinforce a particular social order by asserting as natural what had in fact been created and sustained through deliberate action.

His testimony also acknowledged the growing dissatisfaction that had begun to appear in the early 1960s. Glenn's admission that such a social order "may be undesirable" indicated an awareness that the system he described benefited some while limiting others. Indeed, his presence, and that of the other NASA representatives, before Congress to justify the exclusion of women from the astronaut corps indicated that some cracks had begun to show in the established social order.[33]

In 1962, however, both flying jets and piloting spacecraft remained masculine preserves. Significantly, Jerri Truhill, one of the successful Lovelace candidates, listed the lost jet opportunity as the biggest disappointment of the canceled Pensacola testing, which would have included an airborne EEG performed in a jet. As a pilot, she wanted to fly the most advanced aircraft. As a woman, she wanted to access an opportunity closed to her by the contemporary "social order."[34]

Women pilots who presented themselves as potential astronaut candidates challenged that social order. In 1962 and for years afterward, Tom Wolfe's famous description of the competitive fighter pilot–astronaut ethos as "the right stuff" resonated at NASA. In his tell-all autobiography about life in the hypermasculine world of the U.S. astronauts, Walter Cunningham described how he and the astronauts resisted changes in "our public image, which we enjoyed, as the John Waynes of the space frontier." Women whose flight journals listed commercial ratings and thousands of flight hours threatened that image. And women who had those credentials and whose physical condition equaled the men's fitness challenged the inevitability of the male fighter pilot–astronaut.[35]

"The Most Unconsulted Consultant in Any Government Agency"

Despite their willingness to set aside two days for public hearings on the issue, the women's supporters on the House Space Committee declined to challenge the space agency directly. And July 17 and 18, 1962, the dates of the subcommittee hearings, also represented Chairman Anfuso's last two days on the House Space Committee. Although Anfuso remained sympathetic to the women's cause, he would no longer be in a position to champion it. At noon on the second day of testimony, he abruptly adjourned the hearings, taking the opportunity to announce the end of his tenure on the committee. Cobb and Hart, who had been told that the hearings would run for three days—and who wanted a chance to rebut NASA's arguments—were stunned.[36]

Because she had not been able to present her summation before the entire panel, Cobb submitted a final statement in writing, hoping to reinforce her case and to counter some of the space agency's testimony. She included the detailed results of her medical testing by the Lovelace Foundation arranged side by side with the results from NASA's Mercury astronaut candidates. No amount of new data would change the outcome, however.

The impact of Glenn's presence, only five months after his triumphant flight made him a national hero, cannot be underestimated. Ten months after Gagarin's April 1961 launch, Glenn's single orbit demonstrated that the United States could finally match the Soviet Union in space. The country reacted with jubilation, throwing a ticker tape parade for him. Carpenter's late May flight proved that NASA could repeat the achievement. The astronauts' presence at the hearings especially impressed the representatives. Anfuso, the chairman of the subcommittee, complimented these "great men" and fairly gushed about the astronauts' national example as he observed "women here bring[ing] their children just to look at them." The astronauts even wowed Rep. James E. Fulton of Pittsburgh, who had argued on the women's behalf. He ended his questions on the second day by adding, "In conclusion, may I compliment Col.

John Glenn and Commander Carpenter. We as Americans are very proud of you and the wonderful job you have done, as well as the other astronauts. And I want to tell you, Col. John Glenn is a stellar witness before a congressional committee." Cobb's and Hart's most ardent supporter had been thoroughly dazzled by the astronauts' star power.[37]

NASA's recent space achievements reinforced the space agency's case that it should be left alone to pursue its goals without undue interference. Indeed, the NASA administrator demonstrated his confidence in the power of the agency's recent achievements by holding a well-attended gathering on the afternoon the hearings ended. Three hours after the second day's testimony concluded, NASA hosted an awards ceremony for the pilots of its X-15 rocket-powered hypersonic research aircraft. Notably, George Miller, the chairman of the House Space Committee, who had not attended the subcommittee hearings, found time to appear at this event. Vice President Johnson also attended, spending fifteen minutes alone with Webb before the ceremony. Another key figure in the debate over women as astronauts also accepted her invitation to toast NASA's X-15 jet test pilots. Like Johnson, Cochran enjoyed a private meeting with the NASA administrator.[38]

Recognizing NASA's momentum, the subcommittee's final report deferred to the newly successful agency. Indeed, the title of the subcommittee's report encapsulated its position. Despite taking the revolutionary step of holding public hearings to investigate sex discrimination years before it became recognized as a legal offense, the House subcommittee gave the hearings transcript and the final report a nondescript title—*Qualifications for Astronauts*—that offered no indication of its gendered topic. Instead, the subcommittee followed NASA's emphasis on astronaut requirements. A willingness to advocate for women as women did not yet exist in Congress.

Although the subcommittee's final report included four recommendations, the first one told the Lovelace women and the space agency all they needed to know. The House subcommittee agreed that "NASA should continue to maintain the highest possible level of personal qualifications" in selecting astronauts. The third of the subcommittee's four recommendations did support a women's testing program, but the House subcommittee members hesitated to force any changes that NASA's own representatives had warned might interfere with further space achievements. The subcommittee left it up to NASA to decide when and how to "consider the possibility of conducting a program of scientific research" and training. No such program would be forthcoming.[39]

At an elementary level, Cobb and Hart had been outmaneuvered in this political battle. They fought without influential advocates and against substantial opposition. Because of the short notice—and because the Lovelace women had never met as a group—the rest of the testing participants did not have the opportunity to appear or even to send letters on their own behalf. (Not all the women agreed with Cobb. A few supported Cochran's arguments that women should wait. Others resented the uniform presentation to the subcommittee, given that Cobb had never polled the full group.) Many of the Lovelace candidates now say they wish they had been more organized so they could have pushed harder as a group. They also recognize, however, that better organization would not have changed the ultimate outcome.[40]

The 1962 House subcommittee hearings did not restart Lovelace's Woman in Space Program because the decisions blocking its resumption could not be changed by public hearings. Lovelace and Cochran, whose personal and professional friendship had gotten the Woman in Space Program off the ground, no longer supported restarting the testing. Most significantly, however, space policy makers set priorities in a political and cultural environment that judged women astronauts to be unimportant or even a hindrance to the U.S. space program. National decision makers agreed with the space agency's assessment of women's place in the "social order." Without strong political advocates to challenge the assumptions underlying these priorities, there existed no compelling reason to add women to the astronaut corps that could override the many reasons for excluding them.

Despite these setbacks, Cobb continued to campaign. She hoped that the media attention prompted by the House subcommittee hearings would create a groundswell that might force NASA to change its policies. In her continued correspondence with the space agency, Cobb grew even more persistent, refusing to acknowledge the brush-offs she received from government officials. Because it had become clear that she could not persuade NASA decision makers directly, Cobb began requesting meetings with President Kennedy. The White House referred her letters and telegrams back to NASA.

In the immediate aftermath of the hearings, Cobb also looked for ways to fulfill the requirements NASA had defended before Congress. She pushed the space agency administrator to tell her how she could succeed, "If I went back to college and got an engineering degree and managed some way to get some jet test pilot experience, could you tell me if I'd be acceptable as an astronaut candidate then?" Cobb would not give up if she could possibly persuade the space agency to consider her. "I beg of you, just for the opportunity to prove myself." Webb's refusal to offer any encouragement increasingly irritated her.[41]

By the fall of 1962, Cobb finally stopped trying to get along with the space agency. Rather than appealing for support from NASA, she began speaking publicly about her frustrating relationship with the agency. In a talk at the Air Force Association's Sixteenth National Convention and Aerospace Panorama, Cobb portrayed the female astronaut project as successful until NASA thwarted it. According to her speech, her consulting relationship with NASA illustrated the extent of the space agency's problems regarding female astronauts. As she told her audience, "Since I am not a Ph.D. with three different science degrees, I assumed my [consulting] appointment had something to do with women in space. That was over a year ago and—believe me—I'm the most unconsulted consultant in any government agency today."[42]

Throughout the fall, Cobb continued to criticize NASA publicly. In a *Washington Post* article, she accused the space agency of "bypassing the one scientific space feat we could accomplish now—putting the first woman in orbit." And in a particularly pointed rebuke, she reminded readers that NASA had been chasing Soviet achievements since *Sputnik*: "With that kind of attitude, it's no wonder that we're second in the space race." That same month, she repeated her Air Force Association speech before the Zonta Club of Cleveland. Her complaint about being "the most unconsulted consultant in any government agency today" became a regular refrain in her talks.[43]

Cobb's public denigration of the space agency finally elicited a reaction. In December 1962, a year after NASA administrator Webb first wrote to Cobb suggesting that her consulting arrangement had no future, Webb summoned her back to NASA Headquarters in Washington, DC. He had had enough.

8

Several Epilogues to Lovelace's Woman in Space Program

On December 17, 1962, NASA administrator James E. Webb met with Jerrie Cobb in the presence of Robert P. Young, a NASA staffer who recorded the meeting in detail for the space agency's records. According to Young's memorandum, "Miss Cobb opened the interview by asking Mr. Webb if he had changed his mind with regard to women participating in the program." He had not. NASA's requirements for pilots in the X-15 high-performance jets or for astronauts in its Mercury, Gemini, and Apollo space programs would not change. Webb conceded that the space agency might include scientists on flights someday, but women would still have to meet the same requirements that men did. The NASA administrator had called the meeting not to answer Cobb's challenges but to chastise her for making them.[1]

Webb scolded Cobb and told her to stop referring to herself as NASA's most unconsulted consultant. As he explained it, her one-year consulting contract, which expired in July, had not been renewed. He instructed her to make it clear in her speeches that she no longer served as an adviser to the space agency. Furthermore, Webb questioned Cobb's motivations for continuing to speak out for women astronauts after the House subcommittee hearings had already publicized her views. Given the high stakes of the space program's success, he wondered why she continued to criticize NASA, and he wanted her to stop. Webb had summoned Cobb to Washington to lay down the law about her campaigning.

Toward the end of their meeting, Webb resorted to subtle intimidation, warning Cobb there might be personal repercussions for any further public criticism of NASA. When Cobb retorted that she believed deeply in her actions, Webb suggested that Cobb's reputation would be jeopardized if she kept pursuing her goals in "an irrational manner." If NASA continued to be pressured by her complaints, Webb warned, he would have to tell the media that he never consulted her because he felt she lacked judgment. Young's account of the exchange attempted to soften Webb's thinly veiled threat. "Mr. Webb made this as simply a statement of fact with no hint or suggestion that she take any stand she did not believe in and I'm sure Miss Cobb understood it that way." With this notice of his intentions to fire back in the press if she kept criticizing the space agency, Webb ended his meeting with Cobb. The NASA administrator wanted the woman in space campaign to be over.[2]

Despite the space agency's efforts to make this issue disappear, however, press coverage of the issue of women as astronauts continued throughout 1963. This time the attention did not arise from Cobb's efforts but was prompted by Valentina Tereshkova's flight as the first woman in space. On June 16, 1963, the Soviet Union launched Tereshkova, an amateur parachutist, into a forty-nine-orbit flight on *Vostok* 6 lasting three days. In the United States, the announcement catapulted the question of female astronauts back into the popular press.[3]

"Why Valentina and Not Our Gal?"

Despite her more aggressive stance regarding NASA, Cobb did not manage to capitalize on the media exposure generated by the House subcommittee hearings. Public attention did not equal political support. Throughout 1963, however, Cobb continued to lobby the government for the chance to become the United States' first female astronaut. While Cobb struggled to keep her campaign alive, government officials combated NASA's discriminatory image by promoting the vital role that women played in the space effort.

As the ranking Republican on the Senate space committee, Senator Margaret Chase Smith of Maine spoke as an authority on U.S. space policy. In January 1963 she reminded the audience at a Business and Professional Women's luncheon that astronauts made up only a very small part of the U.S. space program. People needed to remember, she argued, that for every astronaut flying in space, thousands of others had worked to ensure his safety. Among those technical and scientific workers, women counted as fully half of "the unspectacular, the non-glamorous workers—the unsung heroes and heroines." By a

May NBC television interview, Senator Smith had revised the proportions of women at NASA downward to a little over one-third, but her core argument remained the same: women already played a vital part in the U.S. space effort.[4]

NASA also shifted its public relations strategy regarding women. In response to periodic Soviet announcements about plans to fly a female cosmonaut, NASA initiated a two-part approach to public relations. In addition to downplaying the importance of women in space, the space agency began to publicize the women it employed. The connection between the two themes remained qualifications—women's inability to fulfill NASA's minimums for astronaut training and the many talents that female employees already contributed to the space agency.

Robert Voas's February speech at the YMCA provided a good example of the space agency's new two-pronged tactic. As the NASA astronaut training officer, Voas disparaged cosmonaut Popovich's public boasts about plans to fly a female cosmonaut, admitting that he was "rather unimpressed with the technical importance of putting women into space." He cautioned, however, that he did not want the space agency to appear "less chivalrous and less sensitive to the important position of the female sex." Despite his claims of chivalrous intentions, Voas proceeded to mock women's potential as space travelers.[5]

Voas made fun of women's potential physiological advantages, joking that you could save weight by using a female astronaut only if you could persuade her to leave her purse behind. According to the astronaut trainer, the savings in food consumption would work best if space doctors told the female astronaut she looked fat before she embarked on her launch, prompting her to diet during the flight. In addition, since male astronauts had discovered how exciting spaceflight was, Voas argued, women's capacity to withstand isolation no longer mattered. Including a female member on a spaceflight crew to keep men on their best behavior was not necessary. He concluded that flying a woman offered no real benefits, only the risks of launching an unqualified individual.

Although NASA refused to risk women's lives by including unqualified female astronauts, Voas argued, the space agency gladly employed women to fill positions for which they met the established criteria. There were 580 women working at the NASA Manned Spacecraft Center in Houston, Texas. Although he acknowledged that most of them were secretaries, Voas reminded his audience how much any organization relied on its support staff. Voas also cited the female nurses and physiologists who worked one-on-one with the astronauts. Women "calculators" plotted orbital trajectories by hand, and female artists painted the symbols on the Mercury capsules. Two remarkable women even worked as senior NASA scientists in astronomy. (Throughout the 1960s, NASA officials touted the agency's employment of Dr. Nancy Roman, an accomplished female astronomer, as proof of its willingness to hire qualified women.) Furthermore, Voas reminded the audience that no one should underestimate the astronauts' wives. He informed the YMCA group that allowing an unqualified woman to be an astronaut would insult the well-qualified women who already worked for the space agency.

In his conclusion, however, Voas revealed the assumptions about gender roles that underlay NASA's policies. The space agency's astronaut trainer assured his audience, "I think we all look forward to the time when women will be a part of our space flight team for when this time arrives, it will mean that man will really have found a home in space—for the woman is the personification of the home." Not until space had been domesticated—rendered safe and routine, and no longer a front in the cold war—would women participate fully in the space program. In fact, the very presence of women in orbit would indicate that space no longer remained a battlefield for international prestige.[6]

Seeds of the mass movement that challenged such equation of women with domesticity existed in the early 1960s, but they had not yet gained enough strength to take root. Indeed, when NASA officials reacted to the controversy about women in space by publicizing the agency's numerous female employees, they indicated some awareness that they could not afford to dismiss women's concerns outright. In the years before the full flowering of the women's liberation movement challenged the stereotypes that limited women's roles, NASA employed women as secretaries or support staff—or in exceptional cases as scientists—but not as astronauts.

Unlike NASA, Cobb took the Soviet boasts of a female cosmonaut seriously. She feared that if she could not circumvent NASA's unwavering refusals soon, any hopes for beating the Soviets to this space first would be lost. After her unsatisfactory December meeting with Webb, however, she recognized that she was running out of ways to make her case. When her telegrams to the White House generated replies written by NASA officials on the president's behalf, she grew frustrated by her inability to break through the government's unanimous denials. By March she decided to bypass NASA altogether. A direct appeal to the White House might bring the immediacy of the problem to the president's attention. She hoped that a heartfelt petition would convince Kennedy to use his influence to force changes at NASA.

In a remarkable March 13, 1963, letter, Cobb poured her heart out to the president. She hoped that a compassionate staff member who read her message would forward it to Kennedy himself: "It is difficult to write this letter knowing it will be read by your secretaries and assistants and the chances are slim that it will get through to you." She pleaded with Kennedy to sympathize with her views and instruct NASA to reconsider her as a volunteer for America's first woman in space. To back up her arguments, she enclosed a packet of her correspondence with NASA and a scrapbook testifying to her aviation qualifications. She implored the president to evaluate the issue personally: "I have worked, studied and prayed for this over three years now and could not give up without one last, final plea to the commander-in-chief." Her appeals elicited no sympathy.[7]

Once again, the White House referred Cobb's entreaties to the space agency. In April, Webb sent a reply repeating NASA's arguments that no female candidates could meet the space agency's jet test piloting and engineering requirements. Even with a Soviet launch in the works, NASA expressed no interest in competing for this particular space first. Cobb's attempts to persuade the United States to preempt the coming Soviet first kept running into the same dead end.[8]

The lack of concern about the Soviet Union's preparations for a female cosmonaut remained unchanged despite U.S. intelligence confirming the Soviet boasts. More than three months before Tereshkova's flight, covert American sources reported that the upcoming "Soviet space spectacular" might well include a woman in space. According to a Central Intelligence Agency Information Report received in March 1963: "This spectacular event would probably be the launching of a woman into space, or a space rendezvous involving two or three space ships." If American decision makers wanted to match the next Soviet space feat, they had ample warning. Despite reliable reports of Soviet intentions to include a female cosmonaut in their space plans, NASA and U.S. government officials chose not to act.[9]

The CIA's operatives reported both components of the upcoming Soviet space spectacular fairly accurately. While Tereshkova made history as the first woman in space aboard *Vostok 6*, Valeri Bykovsky also orbited the earth in *Vostok 5*. The two spacecraft did not link up or rendezvous, but they did approach each other to within approximately five kilometers. Given that *Vostok 3* and *Vostok 4* had already achieved a near rendezvous in August 1962, however, the presence of a female cosmonaut commanded most of the attention. Despite the public relations success of Tereshkova's flight, however, the Soviet program remained suspicious of female cosmonauts, choosing not to fly any of the other women who had trained with her.

When the Soviet Union began searching for a female cosmonaut, the space program recruited five women to train for the opportunity. Because Vostok space technology relied on the capsule's occupant to parachute from the craft during the descent phase, the Soviet program deemed parachute experience to be as valuable as piloting skills. Two of the women were on the Soviet parachute team. Two belonged to flying clubs, and the last, a rocket propulsion engineer, also flew recreationally. After extensive physical tests and seven months of training, all five women earned military rank and full cosmonaut status. But because probably only one woman would get to fly, the political lobbying on behalf of each candidate intensified as the group waited for a woman's flight to be scheduled. Finally, with enough votes from Soviet space officials and Khrushchev's endorsement, Valentina Tereshkova won the assignment.[10]

The flight of the first woman in space was only half of the June 1963 Soviet space spectacular. The paired flights also set endurance records far beyond any American achievements at the time. Two days before Tereshkova was launched under the code name Seagull, her "space brother" Bykovsky began a flight that set a world space endurance record. Tereshkova was launched two days later, landing after almost three complete days in space. Her time aloft totaled more than all of the United States' previous manned missions put together. The successful twin flights not only achieved international renown for the first woman in space but also demonstrated the Soviets' superior abilities for long-duration space flights.[11]

Within months of her safe landing, Tereshkova provided the Soviet government with another public relations coup. In November 1963 she married cosmonaut Andrian Nikolayev (*Vostok* 3) in a widely publicized state wedding, the first of its kind in the Soviet Union. As much as her successful launch, the marriage provided a triumph for the Soviet space program. The next June, the much-heralded birth of the couple's daughter rounded out the achievements. As a cosmonaut, one half of the nationally acclaimed cosmonaut marriage, and the mother of their celebrated child, Tereshkova became a symbol of Soviet womanhood.[12]

Despite the public relations boons of her flight and her high-profile marriage, biased responses to Tereshkova's in-flight performance damaged her reputation within the Soviet space program. Although there had been no plans to fly another female cosmonaut, space program officials blamed the lack of any subsequent female missions on Tereshkova's substandard functioning in flight. Interviews with Soviet space officials perpetuated rumors of her poor performance. They reported that she succumbed to terrible space sickness, refused to complete her mission objectives, and became hysterical from terror.[13]

Soviet officials assessed Tereshkova's performance hypercritically because she was female. Other Soviet cosmonauts including Gherman Titov, the second cosmonaut in space, experienced space sickness during their flights but continued to fly with the blessing of space program officials. Tereshkova had some problems, but she also met many of the mission goals. Nonetheless, space officials held Tereshkova's frank complaints about flight conditions against her as evidence that a woman could not tolerate space travel well. Despite Khrushchev's assertion of women's full participation in the Soviet state, the USSR never demonstrated a sustained commitment to including women in its space program. The next female cosmonaut did not fly until 1982.[14]

NASA officials bristled at questions about Tereshkova's flight. For example, Wernher von Braun's private correspondence betrayed his growing exasperation with the discussion of women in space. When Martha Ann Wheatley, a *Huntsville (Alabama) Times* reporter, requested von Braun's commentary for a story immediately following the Soviet feat, he could not contain his frustration. In his response to the memo, von Braun noted sarcastically, "Suggest we put Mrs. Wheatley in orbit. Would kill 2 birds with 1 stone." He then declined to "make public comments on this highly touchy subject at this time." Von Braun's private sarcasm reflected other NASA officials' desire to distance themselves from the issue.[15]

Other Western space scientists also dismissed the Soviet flight as scientifically insignificant. When the *New York Herald Tribune* reported the news, the headline questioned the importance of Tereshkova's space record: "It Was a Wonderful Space Feat—but Exactly What Did It Prove?" The article reported that American and British scientists speculated that a rendezvous had been planned but failed. Despite Bykovsky's space endurance record, Sir Bernard Lovell concluded that "apart from the initiation of the space experiment on the girl, the durations of the flight have not significantly extended those of earlier multi-orbit Russian space jaunts." Others did not dismiss the inclusion of "the girl" quite so easily, however.[16]

In the weeks after the Soviet first, newspapers across the country published a wire story by Joy Miller, the Associated Press women's editor. When Miller asked Jerrie Cobb for her response to Tereshkova's feat, frustration tinged Cobb's good wishes for the Soviet woman cosmonaut. Under such titles as "Red Space Girl Irks U.S. Woman Flyer" and "Space-Bitten U.S. Gal Chagrined," Cobb congratulated Tereshkova, but her assessment of American space decision makers was less laudatory. Cobb described talking with NASA's administration as being "like beating your head against a brick wall." Nonetheless, she vowed not to give up. "But I'll keep on beating it, because it's that important to me." In a similar vein, articles with titles such as "Spacewoman up in the Air over Red Tape," highlighted Cobb's campaign for an American woman in space and NASA's reluctance to accept her.[17]

Media attention focused on Cobb not only because of her willingness to speak publicly about the Soviet launch but also because her "as told to" autobiography, *Woman into Space: The Jerrie Cobb Story*, appeared within weeks of the Soviet launch. Written by Jane Rieker, a *Life* magazine reporter who received credit as coauthor, the book dramatized Cobb's life as an award-winning pilot and astronaut hopeful. The narrative hinged on Cobb's performance during the grueling Lovelace Foundation astronaut fitness tests. Ironically, however, that the book's release coincided with Tereshkova's flight highlighted the frustrations plaguing Cobb's crusade. Louise Sweeney used the opportunity created by Tereshkova's flight to review *Woman into Space* in her "Designed for Women" newspaper column. Struck by the coincidence of Tereshkova's launch and Cobb's publication, Sweeney's headline voiced the obvious question: "Why Valentina and Not Our Gal?" Impressed by the accounts of Cobb's performance during rigorous physical examinations, Sweeney endorsed the Oklahoma pilot's campaign for a ride into space.[18]

Tereshkova's flight also brought out other advocates for women in space. Most notably, Clare Boothe Luce, the wife of *Life* magazine magnate Henry Luce and a formidable public figure in her own right, used her position to publish a prominent *Life* magazine article taking American policy makers to task for their inaction. The former congresswoman, ambassador to Italy, magazine editor, and noted playwright had already written about Cobb and women astronauts in a "Monthly Commentary" published in *McCall's* magazine. In that piece she offered a perceptive analysis of the underlying rationale for excluding women from spaceflight. According to Luce: "The most important reason men will not (for a long time) accept women in the astronaut program is that in this field (as in most others) men somehow think their virility and masculinity depend upon establishing and maintaining their intellectual and physical superiority in it." When she learned that the Soviet Union had launched the first woman in space, she took advantage of her connections to *Life* magazine to pen a scathing critique of U.S. decision makers.[19]

Following a three-page spread featuring photographs of Tereshkova preparing for her historic mission, Luce offered a full-page commentary under the acerbic title, "But Some People Simply Never Get the Message." According to Luce, Americans dismissed Tereshkova's achievement as a propaganda stunt because they failed to acknowledge communism's practice of equality between the sexes. She cited statistics demonstrating the numbers of Soviet working women and practicing female doctors. Moreover, she pointed out that flying into space represented more than just another job. The space traveler became "the symbol of the way of life of his nation." By allowing a woman to undertake a dangerous mission laden with international visibility and prestige, Luce stated, "the Soviet Union has given its women unmistakable proof" that it valued them.[20]

In the end, Luce argued, the failure of the United States to match this first signified more than just a deficit in the space race. She charged that American unwillingness even to consider flying a woman reflected a weakness in gender relations that put the United States at a disadvantage compared with the Soviet Union. Given her previous history of staunch anticommunism, that Luce would offer such praise for the Soviet system showed the magnitude of her outrage. She blasted U.S. officials for their lack of vision on the issue. As she wrote, the "failure of American men" to understand the advantages that women's equality gave the Soviet Union "may yet prove to be their costliest Cold War blunder." Luce's *Life* article berated American decision makers for their failure to consider putting a woman in space.[21]

In addition to voicing Luce's critique, however, *Life* magazine's spread also pushed the individual women who passed Lovelace's astronaut tests into the national media spotlight. In a two-page feature titled "The U.S. Team Is Still Warming Up the Bench," *Life* published pictures of all thirteen women who had been accepted to participate in the Pensacola phase. The spread featured an almost full-page photo of Cobb accompanied by smaller pictures of the other twelve with brief biographical captions. The accompanying text pointed out that the United States could have matched the Soviet feat if American decision makers had been willing to fly any one of these women.[22]

As a result of *Life*'s photo spread, the thirteen women from Lovelace's Woman in Space Program suddenly found themselves receiving individual attention as potential astronaut candidates. For example, a feature article about Georgia resident Myrtle "Kay" Cagle that appeared in the *Atlanta Journal* was reprinted in two newspapers in her native North Carolina. Each locality wanted to claim her achievements. Many of the women suddenly enjoyed press attention to their flying careers because of their perceived roles as potential U.S. female astronauts.[23]

Although the *Life* photo spread set off a flurry of publicity for the thirteen women who passed Lovelace's Woman in Space Program tests, the political support that Cobb hoped for never materialized. As a result, the 1962 decision not to resume the Lovelace testing initiative did not change. Lovelace's Woman in Space Program remained defunct. Without any prospects that the testing could be revived, the principal actors in the initiative—Lovelace, Cochran, Cobb, and the rest of the participants—moved on with their lives, pursuing other projects and going in different directions. Public attention dissipated as the key figures either gave up the quest or took their inquiries to private channels. As a result, after 1963 Lovelace, Cochran, and Cobb took paths that explain why the Woman in Space Program disappeared from view during the 1970s and 1980s.

Lovelace and Cochran

Before Dr. Lovelace and his colleagues abandoned the women's astronaut testing entirely, two of the researchers involved in the project, Dr. Johnnie R. Betson Jr. and Dr. Robert R. Secrest, published the group's findings in the February 1964 issue of the *American Journal of Obstetrics and Gynecology*. Their short "rationale and comments"—the only research ever published from Lovelace's Woman in Space Program—allowed the Foundation to publish even if no study worthy of a full journal article would ever be completed. The three-page piece reflected how completely the tide had turned against women as astronauts. The same doctors who administered tests hoping to demonstrate that women offered physiological advantages for spaceflight eventually concluded that women made unreliable astronaut candidates.[24]

The article argued that women's menstrual cycles fundamentally compromised their suitability for outer space. Menstruation complicated not only actual space flight but also the tests needed to qualify any space traveler. To determine whether women could function well in space, Betson and Secrest argued that aerospace physicians would have to conduct "basic physiological work during the entire menstrual cycle." All examinations would need to be repeated several times to determine how women's capabilities differed at various times of the month. Referencing other medical studies, the doctors cautioned that changes in peripheral vision, attentiveness, and coordination had been linked to the onset of a woman's period. In addition, they asserted (with appropriate citations to the underlying articles) that "mental illness is higher, [the] crime rate increases, and there are more attempted and successful suicides just prior to and during the menstrual flow." As a result, women's monthly cycles negated any advantages in size or weight that female astronauts offered.

Since before World War II, similar arguments about menstruation's effects had been used to bar women from flying aircraft during their periods. Difficulties with enforcing such a ban, which required either voluntary self-reporting or an even more indelicate inquiry, coupled with women's proven abilities to fly safely month after month, eventually led to such prohibitions' being ignored. Nonetheless, fallacious claims that women were fundamentally compromised during their periods appeared repeatedly in medical literature. Advocates for women had to disprove these erroneous conclusions about women's physiology over and over.

Even if thorough examinations proved that a female candidate did qualify for space travel, Dr. Betson and Dr. Secrest worried that women's physiology would not prove reliable enough for programmed space launches. In the view of the Lovelace physicians, women introduced complex physiological variables into the already difficult interface between spacecraft and astronaut. They concluded: "Menstruation may complicate the use of the female astronaut in an environment of time tables and rigid schedules needed for a perfectly manned space voyage." With so many perceived disruptions linked to women's changeable physiology, the doctors worried that space programs would have a hard time matching a female astronaut's cycle and a narrow launch window.[25]

In conclusion, Dr. Betson and Dr. Secrest cautioned other researchers that "while any proposed evaluation of these women is interesting and stimulating, serious problems are confronted in progressing with any such venture." In their summary assessment of the Lovelace Woman in Space Program, the researchers, who had begun by hypothesizing that women might offer possible advantages for spaceflight, instead concluded that women were outright unreliable space travelers. Their findings discouraged other researchers from pursuing a female astronaut project for some time.[26]

The dormant state of the Lovelace testing program influenced their conclusions. Although politics did not drive Betson and Secrest to argue against women astronauts, opposition to female space travelers probably affected their willingness to defend women's hypothesized advantages for spaceflight. After all, over two years had passed since any chance remained that more testing might be undertaken. With no prospect of continuing the overall project, there was little incentive to argue for controversial medical judgments. The short "rationale and comment" publicized the effort but ultimately deferred to medical literature that remained suspicious of how menstruation affected women's abilities.

Even though the Woman in Space Program had been Lovelace's personal project, his name did not appear on the article. By 1964 Lovelace had already become deeply involved in other ventures, including working closely with NASA on the upcoming Gemini flights. Then any chance that he might revisit the women's testing project ended abruptly when Lovelace and his wife Mary died in a plane crash on December 12, 1965.

Dr. and Mrs. Lovelace perished along with Milton Brown, a twenty-seven-year-old charter pilot, when their twin-engine Beechcraft Travelair crashed in a high mountain valley on a return flight from Aspen, Colorado. Brown had flown the Lovelaces to visit their new home near Aspen and to visit a member of the Lovelace Foundation's board of directors in Colorado Springs. Some time after he pointed the airplane south toward Albuquerque, however, he mistook one mountain valley for another and flew into a dead-end box canyon. He attempted to turn the airplane around, but its wing clipped the canyon wall. The aircraft crashed into the snow-filled valley, killing the Lovelaces on impact.[27]

The news of Dr. Lovelace's death devastated his family, friends, collaborators, and coworkers. The Lovelaces' three daughters had been orphaned. As a friend of the couple and the godmother to Jacqueline Lovelace, Cochran mourned his death deeply. Uncle Doc, Randy's eighty-two-year-old uncle and medical partner, never fully recovered. At the Lovelace Clinic and the Lovelace Foundation for Medical Education and Research, the sudden tragedy stunned Randy's colleagues. They had lost a visionary leader, a skilled researcher, and a respected friend. In many ways the accident's aftershocks reverberated throughout Lovelace's namesake research center throughout the 1970s and 1980s as his successors tried to continue his work in a rapidly changing medical environment.[28]

The U.S. space effort also felt the absence of Dr. Lovelace, who had been named NASA's director of space medicine in April 1964. When searchers reached the snowbound crash site on the same day that *Gemini 6* and *Gemini 7* docked in space (the first orbital rendezvous of two spacecraft), President Lyndon Johnson paid tribute to Lovelace during his public comments. Three months after the fatal accident, Dr. George E. Mueller, NASA's director of manned spaceflight operations, read a seven-page eulogy into the official record of the Senate Aeronautical and Space Sciences Committee. Senator Clinton P. Anderson, the committee's chairman and Lovelace's friend, also spoke. Mueller's eulogy, like many other obituaries, cited Lovelace's record-setting 1942 parachute jump as an example of the doctor's characteristic approach to aerospace medicine. The Senate committee tribute (and newspaper accounts of Mueller's eulogy) even reprinted the War Department's eloquent report of the jump in its entirety.[29]

Lovelace left a legacy characterized by bravery, innovation, and vision. World War II pilots owed him a debt for the BLB oxygen mask's lifesaving role in high-altitude flights. His medical colleagues credited him for pushing the boundaries of aviation medicine until it became space medicine. Government collaborators from the air force and NASA remembered him for his work on projects including the U-2 spy planes and Project Mercury. His absence represented a tragic loss not only for family, friends, and colleagues but also for the Woman in Space Program.

Because of his other research interests, Lovelace had not advocated for resuming the Pensacola testing in 1961 beyond sending one well-argued letter to Webb. His premature death meant he never had the chance to revisit that question. Who knows how much sooner women might have been introduced into the astronaut corps if this dynamic man who had measured their abilities personally had remained present to talk about his experiments? Lovelace never had the opportunity to translate his firsthand experience into renewed actions when the climate for women's participation warmed. Such speculation about reviving the female astronaut project is warranted because Lovelace's financial partner stayed interested in women astronauts for years after 1962.

Despite testifying before the House subcommittee that women should not be injected into NASA's newly successful program, Jacqueline Cochran did not give up on leading a female astronaut project. Within weeks of arguing that running a women's space program would waste government money because candidates might abandon expensive training for marriage and children, Cochran contacted NASA to propose a large-scale female astronaut testing program of her own. In fact, when Cochran visited Washington, DC, for the July

1962 House subcommittee hearings, she carried a draft memorandum addressed to the NASA administrator and deputy administrator. When she returned to her office, she revised the memo and sent it to Webb and Dryden, proposing large-scale research into women's capabilities for space.

Cochran's plan called for thoroughly testing "not less than 100" women, a project she guessed "would take well over a year to complete." Although she planned to include the women who had passed Lovelace's tests, she proposed recruiting the rest of the female volunteers from the armed forces. With military participation, she argued, enlisted women could be examined at several sites simultaneously, reducing the time needed. On completing this research, Cochran suggested, NASA could consider whether further examinations in jet aircraft or space simulators would be worthwhile. Before initiating any real women's space program, Cochran believed, the space agency needed additional research that would require hundreds of volunteer subjects and possibly years to complete.[30]

Although NASA did not act on her recommendations, Webb rewarded Cochran for her support during the House subcommittee hearings by naming her a NASA consultant on June 11, 1963, five days before Tereshkova's flight. Cochran's support of the space agency during the public hearings and her continued correspondence with NASA on the subject endeared her to the agency's administration. As deputy administrator Dryden assured Cochran: "As I think I have said to you before, you have taken a very statesmanlike view of the role of women in the national space program." Although the appointment may have been in the works for some time, its timing seems curious. Rumors of a female cosmonaut had put women space travelers back on NASA's radar. Also, quite uncharacteristically, Cochran chose not to have her appointment publicized. Just five days before Tereshkova entered orbit, Cochran took her oath as a NASA consultant without much publicity. Regardless of its quiet beginning, Cochran took great pride in the consulting arrangement with the U.S. space agency, maintaining her official NASA connection for over eight years.[31]

Cochran repeated her offer to head an exploratory female astronaut testing program several times during her tenure as a NASA consultant. On her appointment in 1963, she reminded Webb of her House subcommittee testimony and reacquainted him with her plans for a "medical research project" on women. The next year, Cochran contacted Lovelace to feel out his position on restarting the Woman in Space Program. She wanted his assessment of how his NASA affiliation could aid the project: "Are you now in a position to give it a real boost and has the time come for me to submit such a project to NASA directly or through you?" Cochran supported testing women as potential astronauts as long as she could lead the program.[32]

By 1967 Cochran had scaled down her proposal in an effort to persuade NASA that a female astronaut research project could be "modest, methodical and thorough-going." Specifically, she suggested: "*Maybe* the time has come when some sort of a preliminary program should be started connected with women in space. If so, I think I can be of great service to my country once more in heading it up." Despite Cochran's 1962 testimony that women might not make good subjects for expensive training, her real concerns had been the size and perceived leadership of Lovelace's Woman in Space Program.[33]

Although the space agency never adopted her proposals for female astronauts, Cochran enjoyed her status as a NASA consultant, taking advantage of the opportunities it offered to be present for important space events. She received invitations to attend the launches of *Gemini* 6 and *Gemini* 12. She attended NASA luncheons and received copies of a new NASA photograph of the surface of Mars. When the White House hosted a black tie dinner in December 1968 to honor James Webb and the Apollo astronauts, she made the exclusive list of invitees. Cochran's attendance at NASA-sponsored events allowed her to maintain contact with prominent space decision makers while also demonstrating her own status. Because Cochran prided herself on her NASA affiliation, she worked hard to maintain it.[34]

At least in part, Cochran's longevity as a consultant stemmed from her refusal to resign. When the space agency made the mistake of ending her appointment with a one-sentence form letter in 1969 ("enclosing herewith the Standard Form 50 which officially terminates your consultancy with NASA"), Cochran took NASA administrator Thomas O. Paine to task. She shot back a two-page missive. Cochran listed all the trips she had taken on NASA's behalf, citing her personal investment of time and money on behalf of the space agency. "I am out of pocket at least $5,000, plus considerable personal time in trying to promote favorable opinion of our space program." In particular, however, Cochran reminded Paine of how her presence before the House subcommittee in 1962 had helped the space agency. She pointedly reminded the new NASA administrator that "Dr. Dryden and Mr. Webb were very appreciative of the position I took." NASA promptly renewed Cochran's appointment as an official consultant. Even seven years after the fact, Cochran saw her subcommittee testimony as a significant point of alliance with NASA.[35]

Throughout the 1960s and 1970s, Cochran remained simultaneously brilliant, forceful, well connected, and difficult. She maintained powerful friendships, hosting presidents and generals at her ranch in Indio, California. She also continued to be a prominent force in aviation and space. As with the Lovelace program, however, her actions did not always benefit women generally. She spoke out actively against admitting women to the Air Force Academy.

Despite her formidable talents and powerful friendships, Cochran's strident personality occasionally hindered her ambitions. When her old friend Lyndon Johnson began his first full term as president of the United States, she volunteered her services for his administration. Unfortunately for her prospects as a political appointee, at some point she had clashed with John W. Macy, the staff member charged with assessing her potential. In 1965 he summarized her reputation in an internal memorandum: "Those who have worked with Miss Cochran tend to be somewhat reluctant to renew such associations." Because of her connections to the president, however, the staff kept her résumé on file. But when a potential appointment arose in 1967, Macy disparaged Cochran's style more pointedly: "She is an insufferable egotist and is very difficult to get along with." Those who disliked Cochran resented her headstrong determination; those who liked her tolerated it because of the extraordinary friendship it accompanied. Even Donald E. Kilgore, Cochran's great admirer and personal physician during her last years, recalled affectionately, "She was one hell of a difficult patient."[36]

Throughout the 1970s, Dr. Kilgore saw more and more of his difficult patient as her chronic health problems worsened. Whenever Cochran did not feel well, she sought help at the medical institution she trusted most: the Lovelace Clinic. She also continued to mourn Randy's loss. As Kilgore recalled, "When I say Lovelace, I mean the institution. When she said Lovelace, she meant the man, because she was just sure that Randy Lovelace was the future. And when he got killed, it was just an unacceptable tragedy to Jackie and so many other people." The medical institution that her friend created—and that she and her husband had helped fund—remained the only one she would patronize.[37]

Until her death in 1980, Cochran maintained a close, if eccentric, connection with the Lovelace institution. Although Cochran trusted the Lovelace physicians, she refused to be treated as a regular patient. She arrived in a motor home because she refused to stay in a hospital room. Instead she hooked the vehicle up behind the Lovelace buildings and lived there during her treatments. As the 1970s wore on, her stays in Albuquerque grew more frequent.

Troublesome health issues had dogged Cochran throughout her life. In many ways, she accomplished her physically demanding aviation records in spite of her physical condition rather than because of it. When she reached her seventies, her body became less quick to recover from her ailments. Finally, after numerous visits to the Lovelace Clinic for her ill health, Cochran died on August 9, 1980. At her death, she held more speed, altitude, and distance records than any other pilot, male or female. Her life as an internationally famous pilot, businesswoman, and celebrity was an extraordinary story.

Cochran titled her 1954 autobiography *The Stars at Noon,* a reference to flying so high that the blue sky faded and the stars appeared—flying to the edge of space. For years, Cochran held on to the idea of leading women into space as she had led women pilots into military flying during World War II. In the end, however, after Lovelace's initiative ended, she could not persuade NASA to organize a new women's testing program. Despite Cochran's repeated proposals, examinations of women for space fitness did not resume in the 1960s.

Jerrie Cobb and the FLATs

Like Cochran, Cobb lobbied for a women's astronaut testing program for several years after the 1962 House subcommittee hearings. A month after Tereshkova's historic 1963 flight, NASA publicly rejected Cobb's formal application to join its astronaut training program. As it had for Project Mercury and the Gemini Program, the space agency recruited astronauts for the Apollo moon missions in groups. On July 12, 1963, Cobb applied to be considered for the third astronaut class. Unfortunately, the July 1 deadline for applications had already passed. A NASA spokesman confirmed that the agency did not include her application in the formal pool because she missed the cutoff date.[38]

Two other women did have their applications accepted, submitting their materials in plenty of time. Nonetheless, they never became astronauts. Space agency officials maintained that they evaluated potential candidates based on qualifications alone, with no regard for race or gender. In a less guarded moment, however, Robert Gilruth, the head of NASA's Manned Spacecraft Center, offered the Copley News Service a more candid assessment of women's applications. According to Gilruth, "Although they [NASA astronaut qualifications] do not specify that applicants must be male, this is the case. NASA does not expect that any women will meet the qualifications or will be selected." He was right. NASA's third astronaut class remained all male (and all white). As the agency prepared for the Apollo moon missions, NASA had no room in its mission objectives for acting as an agent of social change.[39]

In 1964 Cobb scaled back her requests to the government. She inquired whether the White House, Congress, or NASA would consider allowing a woman to pilot an experimental X-15 research jet to the edge of space (approximately fifty miles). Such a feat would put the United States in the record books for having flown a woman in space, even if she did not make an orbital flight or even achieve a rocket launch. They said no. Cobb would not be riding a rocket. She could not get a flight on the X-15. She would not get to go into space at all.[40]

The slow realization that her chance at spaceflight had evaporated hit Cobb hard. At some point in 1965, she finally accepted that she would never persuade government officials to allow a woman to fly into space. No amount of faith or effort could change NASA's policies. She withdrew to a Jamaican village for a month to contemplate her future.

During her retreat, she returned to an idea that she had considered before she ever thought about outer space: being a missionary. With the undeveloped regions of the Amazon drawing her, Cobb set out to become a flying humanitarian. When she approached established missionary groups, however, the organizations expected trained missionaries who had learned to fly, not experienced pilots who wanted to serve. They certainly did not expect a woman. Cobb decided to strike out on her own. With her parents' assistance, she purchased her own airplane and flew south into the Amazon basin. The end of Lovelace's Woman in Space Program marked the beginning of a new mission in Cobb's life, a vocation as a flying emissary to the indigenous peoples of the Amazon.[41]

Cobb carried out her missions in a refurbished twin-engine Aero Commander aircraft that she dubbed *Juliet* and that the Amazon natives called "the bird." During her years in Amazonia, she updated her financial supporters about the good work their donations permitted through an occasional newsletter describing her latest actions. She wrote about the villages that kept primitive flying strips clear so she could reach them. The challenges of keeping these runways open—and of taking off and landing on them—increased each year during the rainy season. Nonetheless, flying provided the mainstay of her work. She told her stateside friends about bringing seeds and rice into the jungle to replace disappearing native foods. She transported nurses in, and when there was a medical emergency she flew sick people out. Throughout these ventures, she assured her supporters, she reminded her newfound friends of God's eternal love.

Back in the United States, the press also reported news from Cobb that went beyond traditional missionary duties. When a Peruvian airliner carrying ninety-two people crashed in the jungle on Christmas Eve 1971, she spent days searching the area from dawn until dusk. Her actions paid off when she found the accident scene on the twelfth day. Ninety-one people had died, but one sixteen-year-old girl flew back out of the jungle in Cobb's airplane. Cobb also reported that she acted as a mediator between pharmaceutical or petroleum corporations and the peoples she had gotten to know, trying to lessen the consequences of modern development for her indigenous friends.[42]

Over the years, Cobb's work elicited both praise and skepticism. In 1972 Cobb's humanitarian flights put her name in the annals of aviation's top honors once again. The Harmon International Aviation Trust awarded her its Harmon Aviatrix Trophy for her flights in the Amazon region with her light plane. At a White House ceremony in September 1973, President Nixon presented the prestigious award, which dubbed its recipient the "world's best woman pilot." She was also nominated for the Nobel Peace Prize. In 1983, however, a *Miami Herald* reporter questioned whether Cobb's missionary zeal had caused her to exaggerate her altruistic successes to her supporters.[43]

Since the mid-1960s, Cobb has spent almost four decades flying in the Amazon. To support her mission, she established the Jerrie Cobb Foundation based in Coral Gables and later in Sun City Center, Florida. She used the Florida base to rest, restock, and refuel for her long forays into the rain forest. Cobb has maintained her dedication to the indigenous peoples of the Amazon region, a calling that emerged in the painful aftermath of her campaign for space.

Although some of Cobb's "fellow lady astronaut trainees" later contributed to her missions, most left the Woman in Space Program behind soon after the telegrams canceling the Pensacola plans arrived in 1961. Most of the women did not learn of the House subcommittee hearings until they were already in session. By the time *Life* magazine publicized the women's pictures in July 1963, their firsthand participation in the campaign for women astronauts was already almost two years behind them. As a result, the transition to everyday life was less traumatic for most of them than it had been for Cobb. Those who had forfeited jobs to take the examinations scrambled to find new places to work. Others simply returned to their existing jobs, families, and flying plans. Over the next forty years, their lives took very different paths.

The women who gave up jobs to participate in Lovelace's Woman in Space Program faced differing prospects. When Irene Leverton got home, she found herself demoted. After looking for better work, she flew as a corporate pilot for three companies and for the U.S. Forest Service on contract. When Sarah Gorelick learned about the Pensacola cancellation, she also struggled to find a job (especially one that would use her mathematics degree as her former AT&T position had done). Finally she took a job in a family business, doing bookkeeping and retail sales. For a young woman in love with flying, work financed her real passion: aviation. Quitting her job forced Gene Nora Stumbough to leave a comfortable situation. Without that push, she would not have become a demonstration pilot for Beech aircraft. As one of a team of three pilots, she flew one of three new Musketeer aircraft around the country in formation. It was an uncommon opportunity for any pilot, and especially for a woman.[44]

After the Pensacola cancellation, most of the women pilots returned to flying either recreationally, competitively, or professionally. As one of a small number of women with an air transport license, Jan Dietrich used her credentials to further her aviation career well into the 1970s. Myrtle "Kay" Cagle returned to her work as both a pilot and an airframe and powerplant (A&P) airplane mechanic. She still works on airplanes in various shops near her Macon, Georgia, home and likes to fly old-fashioned "tail-draggers" (airplanes with a wheel on the tail instead of a "tricycle" landing gear under the cabin).

Several of the women maintained their existing aviation businesses or developed new ones. Jerri Sloan returned to Texas, married her business partner Joe Truhill, and helped to raise his son as they continued their aviation business. Bernice "B" Steadman and her new husband maintained Trimble Aviation, the business she had started. They continue to be business owners, running a taxi company in Traverse City, Michigan. After her tour with the "Three Musketeers," Gene Nora Stumbough married Robert Jessen, and they started a Beech aircraft dealership in Boise, Idaho. After retiring from selling aviation insurance and running a fixed base operation (private aviation center), Jessen published a history of the 1929 Powder Puff Derby.[45]

Not all of the women kept their primary identifications as female pilots. Marion Dietrich continued to combine her flying with her career as a writer. Jean Hixson worked as a fifth-grade teacher while she flew in the Air Force Reserve. Sarah Gorelick Ratley married, had a daughter, and continued to fly recreationally but worked as an accountant during the day. In 1991 she became a certified public accountant. At least one of the Lovelace women gave up flying entirely. When Rhea Hurrle met a wonderful man who asked her to marry him and to quit flying, she did both happily. The new Mrs. Woltman became a professional parliamentarian. Whenever she flies now, she takes commercial flights.[46]

With two notable exceptions, few of the women who participated in Lovelace's Woman in Space Program took part in breaking down other gender barriers. These fliers accepted Lovelace's extraordinary opportunity because they wanted to push their limits as pilots, not to lobby for women's rights. When the women's liberation movement developed, most chose not to identify as feminists. Jane Hart was the first exception. Reading Betty Friedan's *Feminine Mystique* got her excited. Because Hart was an acquaintance of Friedan's, the author asked her to be a founding board member of the National Organization for Women (NOW) in 1966. Because her husband's Senate career split their residence, Hart founded chapters in both Michigan and Washington, DC. Since her husband's death, she manages a scholarship fund in his name and travels widely, visiting her eight children and many grandchildren.[47]

Wally Funk was the second woman from the Lovelace group to break down gender barriers when she helped to advance women's presence in the field of aviation safety. In the 1970s, as American women pushed for their places in the public world, aviation slowly began to accept women in roles beyond the Powder Puff Derby. Funk became one of the first female inspectors for the National Transportation Safety Board. Today she continues to lecture widely on aviation safety in a dynamic presentation based on her experiences with the NTSB.

Since the end of the Woman in Space Program, two of the thirteen women who passed the examinations have died. After a long career spanning the WASPs, the Lovelace program, and the Air Force Reserve, Akron schoolteacher Jean Hixson succumbed to cancer in 1984. Her sister, Pauline Vincent, maintains Hixson's memory, reminding people about her sister's contributions to aviation history. Marion Dietrich has also died. Her twin, Jan Dietrich, survives in San Francisco. Their family keeps the memory of their flying achievements alive in a scrapbook of clippings and mementos. Mindful of the twins' significant achievements, their sister Pat Daly honors their remarkable place in her family.[48]

In the decade immediately after the Pensacola cancellation, the story of Lovelace's innovative women's testing program remained a largely private history, shared within families or by the community of women's aviation.

Throughout the 1970s, only one article brought the program to national attention. In 1973 feminist newspaperwoman Joan McCullough rediscovered "The 13 Who Were Left Behind," publishing her findings in the September 1973 issue of *Ms.* magazine. Her inaccurate introduction accused NASA of hiding the fact that the space agency had put women through "the initial Mercury Astronaut Candidate Testing Program."

Writing for the one-year-old magazine, McCullough used the story to expose women's continued exclusion from the front lines of the space program. Even after NASA turned from the Apollo Program's focus on exploration to a new emphasis on research, women still did not have a place in the astronaut corps. According to the article, the discrimination that the thirteen women experienced in the 1960s represented the first episode in a long history of NASA's dismissal of women. As McCullough informed her readers, the space agency continued to reject well-credentialed women: "In 1967, 17 women with advanced degrees in fields directly related to space were among those reviewed by the 900-member all-male science selection panel. The women were all bypassed for 11 men, including four men in their twenties and four others who hadn't yet obtained doctorates."[49]

Until 1978, NASA rejected all women who applied for astronaut training. According to the space agency, the selection committee reviewed all applications impartially, considering only NASA's astronaut regulations without taking any notice of race or gender. Nonetheless, the astronaut corps remained all white and all male. When the space agency finally selected female and nonwhite astronaut candidates, they admitted the new recruits in a newly created two-track system of selecting astronauts for the shuttle program.

To conduct the orbital experiments made possible by NASA's new Space Transportation System (STS) or space shuttle, the space agency created a new kind of astronaut. These "mission specialists"—scientists, physicians, and researchers—conducted experiments while a separate group of pilot astronauts commanded and flew the missions. All the women admitted into NASA's astronaut program in the 1978 class entered as mission specialists. These women held PhDs and MDs. None were pilots. As a result, when mission specialist Sally Ride became the first American woman in space in 1983, her achievement did not excite the women who had participated in Lovelace's testing program. They did not identify with scientists who held advanced academic degrees. Not until NASA announced the first woman pilot astronaut did they feel that their dreams for American women in space had finally been fulfilled.

"Send Jerrie into Space"

NASA astronaut Lt. Col. Eileen Collins first learned about the women of the Lovelace testing program when Gene Nora Jessen, one of the successful Lovelace candidates—and the president of the Ninety-Nines women's aviation organization during the early 1990s—became interested in bringing the group together for a documentary. Working with a filmmaker who had learned about the Lovelace testing while researching another women's aviation project, they invited the women to gather at the Ninety-Nines Headquarters in Oklahoma City, Oklahoma. They also invited Collins, a former air force pilot and fellow Ninety-Nine. When she met the women in Oklahoma City, she recognized how their story anticipated the history she was about to make as NASA's first female pilot astronaut. Through women's aviation circles, she had heard something about the 1960s testing project but, as she recalled, "I didn't know that these women were out there and that I could just go talk to them and learn about what they had done." The meeting established an immediate bond between Collins and the women who had dreamed of qualifying for her job thirty years earlier.[50]

The "reunion" also helped to create a collective identity for the Lovelace women. Because the Pensacola arrangements dissolved before they arrived in Florida, those who passed the Albuquerque examinations never assembled in one place during the 1960s. Even though many of the pilots knew each other through women's aviation connections, few realized exactly who else had made the final list of thirteen Pensacola candidates. Furthermore, a complete accounting of all twenty-five women invited to take the Lovelace tests remained lost until 1997. Because two of the successful test subjects, Marion Dietrich and Jean Hixson, died before the first reunion, the entire group of thirteen successful Woman in Space Program candidates never gathered in one place. In a series of gatherings held in the 1990s, however, the women became known as a group, linked by their common experience of taking Lovelace's astronaut fitness tests. They became known as the "First Lady Astronaut Trainees" or FLATs, adapted from Cobb's 1961 salutation, "Dear F.L.A.T. (fellow lady astronaut trainee)." Throughout the 1990s, renewed public attention to their collective history increased the sense of the women as a group.[51]

Despite this new collective identity, the women's recollections of their expectations—and therefore, of the greater meaning of the Lovelace testing episode in their lives—differed significantly. For some, the examinations represented a brief seven days of aerospace tests, notable but dwarfed by their later aviation careers and family lives. Others remembered being promised at least a fair chance of being vetted for space fitness, an opportunity that vanished because men simply could not envision women as competent space travelers. Such differing memories resulted naturally from a testing program that required the participants to take part individually rather than as a unified group.

As a group, however, the women finally received some closure for this unfinished chapter in their lives when Collins invited them to witness her 1995 launch on STS-63. The experience thrilled them. As Collins recalled in 1999, "I invited them to all three of my launches. And they came. Jerrie Cobb came to all of them. I actually flew something for her on my second flight." Collins carried small mementos into space for several of the women. For some, however, just being there was enough.[52]

For at least one of the Lovelace test subjects, the connection with Collins helped to keep alive her dream of going into space. For Wally Funk, knowing Collins allowed her to make some tangible progress toward her continued goal of flying in space. Funk considers the Ninety-Nines pin that Collins flew on a shuttle mission for her to be her "first payload." Funk has traveled to Russia twice (in 2000 and 2002) to undertake the space training offered to paying customers at the Star City cosmonaut center. She also signed on as the space pilot for Interorbital Systems, a California-based firm seeking to compete for Peter H. Diamandis's $10 million X Prize (to be awarded for a successful, repeatable, privately funded human spaceflight with at least three passengers). She plans to fly Interorbital's Solaris-X suborbital manned rocket plane. She is also affiliated with Trans Lunar Research, a commercial firm aiming to set up the first manned lunar station on the surface of the moon. One way or another, Funk plans to fly in space.[53]

Collins's connection with the Lovelace women not only inspired the women themselves but also publicized their history. During their attendance at Collins's 1995 launch, the women pilots suddenly found themselves subjected to a barrage of media attention. Their story appeared on television, in newspapers, and on the radio. After the initial meeting with Collins at the Ninety-Nines Headquarters, all the surviving women from Lovelace's Woman in Space Program gathered for the first and only time at the Smithsonian Institution's National Air and Space Museum to film a *Dateline NBC* segment, broadcast on February 10, 1995. In the process, the group gained the moniker the "Mercury Thirteen," in parallel with the original seven Project Mercury astronauts, who were called the "Mercury Seven." Newspapers and radio carried the news as a human-interest story. After Collins invited the Lovelace women to attend her second launch, John F. Kennedy Jr.'s short-lived chronicle of modern political culture, *George* magazine, featured the women's history in its glossy pages as "Rocket Grrrls!" Lovelace's Woman in Space Program was suddenly all the rage.[54]

In a social and political environment sensitized to women's past discrimination, the story of a canceled women's astronaut testing program in the early 1960s touched a nerve. A media-savvy public, unsurprised by past gender bias and eager to be indignant on the women's behalf, relished the tale. Furthermore, the program's hazy beginnings and vague reasons for cancellation created places for erroneous versions of the history to lodge. Newspaper articles and magazine features repeated inaccuracies that amplified the women's rejection, casting the Lovelace women as astronaut candidates initially recruited by NASA only to be cast aside later. The misrepresentation of the circumstances of the Lovelace women gradually eclipsed the more subtle story of the actual difficulties they faced. More insidiously, the exaggerations undercut Cobb's second attempt to persuade NASA to grant her a flight into space.

When NASA administrator Daniel Goldin announced that former Mercury astronaut and sitting senator John Glenn would return to space aboard STS-95 in 1998, he inadvertently resurrected the question whether Cobb should be allowed to become an astronaut. To many observers, NASA's decision to send the United States' first orbiting astronaut back into space appeared to be a political payback for the senator's support of the Clinton administration. During difficult congressional hearings, including the Whitewater investigations, the Ohio senator had been a valuable ally to the administration. Glenn let it be known within the Beltway that the only thing he really wanted that was not within his own power was to return to space. In January 1998, before a NASA Headquarters auditorium packed with television cameras and reporters, Goldin announced Glenn's flight assignment to STS-95. Despite the political undertones of the septuagenarian's return to active flight status, however, the NASA administrator emphasized that the aging astronaut's desire to investigate geriatric medical concerns had been the deciding factor. Striking similarities existed between spaceflight's taxing consequences for the human body and aging's effects on bone and tissues. To those skeptical of Glenn's return to space, however, NASA's justification about running a geriatric medical study featuring a single male subject rang hollow.

Constance Penley, a feminist scholar and author of *NASA/TREK*, a gender analysis of NASA and American culture, called for the agency to balance its Glenn decision by including one of the women from Lovelace's Woman in Space Program on a shuttle flight. Friends and fans of Jerrie Cobb had the same thought. Those who knew her history—and who remembered Glenn's pivotal role at the 1962 House subcommittee hearings—had listened to rumors about his potential return to space with chagrin. After the announcement, Cobb's proponents began talking actively about appealing to NASA on her behalf.[55]

The campaign to send Cobb into space gained momentum in early March 1998 when Jean Ross Howard Phelan, the head of the Whirly-Girls women helicopter pilots' association, talked to then first lady Hillary Rodham Clinton and NASA administrator Daniel Goldin during the White House press conference announcing Eileen Collins's assignment as the first woman to command a space shuttle. At that occasion, Phelan presented the first lady and the NASA administrator with copies of Cobb's newly published autobiography, *Jerrie Cobb: Solo Pilot.* She reported to Cobb's close friends that the first lady seemed interested enough in the story to make some inquiries about it.[56]

In the weeks that followed, the campaign for Cobb to take a shuttle flight picked up speed. In March, NOW circulated a petition at the 1998 Women in Aviation, International, annual meeting in Denver, Colorado. The text asked petitioners to attest to their support for Cobb's flight. Signers received a lapel sticker urging, "NASA: Send Jerrie into Space." In support of the campaign, Cobb returned from the Amazon to appear at the conference. Sponsored by the Whirly-Girls, she stationed herself at their exhibit booth, selling copies of *Solo Pilot* and talking excitedly about the possibility that her quest for space might not be over.

Even after thirty-five years, however, the pain of her history with NASA remained deep. In fact, Cobb's memories of the early 1960s remained so tormenting that she had not written anything in *Solo Pilot* about her first battle for space. The new autobiography covered the early 1960s by filling a short chapter with photographic reproductions of newspaper clippings. Nonetheless, the allure of going into space remained so strong that even after decades of living as a missionary, when the memories of her first battle for spaceflight still remained too painful to address directly, Cobb jumped at the chance to resurrect her hopes. Her supporters encouraged people to send letters to NASA urging the agency to schedule Cobb for a shuttle flight. After all, the precedent had been set.[57]

NASA decision makers had already been swayed by persistent letter writers. At the same press conference in which Goldin proclaimed Glenn's return to space, the NASA administrator also announced Barbara Morgan's readmission to the astronaut corps. Morgan, Christa McAuliffe's backup, had been grounded when NASA put the teacher-in-space program on hold after the space shuttle *Challenger* exploded on January 28, 1986. In response to a ten-year letter-writing campaign, however, NASA officials reconsidered, reactivating Morgan's flight status. Clearly, determined public pressure could sway the space agency's determination that only scientific and technological factors influenced crew composition. After Morgan's reappointment proved that NASA's astronaut policies had loopholes, Cobb's supporters aimed to open another one.

By July, Cobb's cause began receiving national attention. *USA Today* reported that in addition to NOW's petition, Donald Dorough, a Fresno Pacific University education instructor, had organized a letter-writing campaign on her behalf. Once again, NASA administrators found themselves facing the question whether Cobb should be allowed to join its astronaut training. Her supporters flooded the space agency with letters and petitions. Sympathetic news stories further fueled the campaign. The Associated Press news wire carried a favorable article by aerospace writer Marcia Dunn to newspapers across the country.[58]

Faced with repeated inquiries from the public and the press, the space agency's public relations officers struggled to respond to allegations of past NASA discrimination. They scrambled to get the history of the Cobb's testing straight so they could answer the increasing media inquiries. The NASA Headquarters History Office provided background information. Unfortunately, since many of the reporters who called the space agency repeated one of the erroneous revisions of the history, NASA's public relations department deflected criticism by correcting reporters who asked whether the Lovelace testing initiative had ever been a NASA project. Few journalists knew enough about the real history of Lovelace's Woman in Space Program to ask questions that would have revealed NASA's role in the testing's demise or the significant discrimination that Cobb did face in the early 1960s.

NASA decision makers handled the 1998 campaign much as the agency had treated Cobb's campaign in the early 1960s: as a public relations problem. The issue never reached the administration or the space flight department, which held the power to make real decisions about scheduling flight crews. In the highly sensitized political environment of the late 1990s, NASA officials trod very carefully to avoid dwelling on past discrimination. The agency also wanted to avoid getting into a situation in which NASA administrator Goldin would have to turn down Cobb's application publicly. As in the early 1960s—without sufficient political support and in the face of significant NASA reluctance—the appeals to "Send Jerrie into Space" did not yield results.[59]

Unlike the 1960s, when the issue faded relatively quickly, this time people latched on to the compelling history of Lovelace's Woman in Space Program. In many ways, the story provided a classic tale of reaching for a dream and hoping against the odds. That women acted as the central figures only added to its appeal. In the 1990s, various artists drew on the history as an inspiration for art, theater, and film. British photographer Nickie Humphries reenvisioned the Lovelace women pilots as the subject of a 1998 photographic essay. In October and November 1999, the New York City Six Figures Theatre Company presented Kate Aspengren's play *Flyer*, loosely based on Jerrie Cobb's life and the story of the "Mercury Thirteen." Various production companies also pursued the rights to the Lovelace women's story with plans for television dramatizations and major motion pictures.[60]

Since the 1990s, the story of Lovelace's Woman in Space Program has once again enjoyed a revival. Many who learn about the program find themselves, like Eileen Collins, surprised that the subjects of this historical battle for women's place in space are still available, that you can go right up and talk to them. As a result of the publicity created by their attendance at Collins's launches, they have become popular symbols, readily received by a public willing to sympathize with women's past struggles and eager to hail the group as ahead of its time. If anything, however, Lovelace's Woman in Space Program means much more because it was so completely a product of its time. The testing initiative that Lovelace created opened a brief opportunity for women to envision themselves going into space, not as science fiction characters but as astronauts chosen because of their piloting skills and physiological advantages. In the end, the research project collapsed, its end overdetermined by the powerful political, social, and cultural forces aligned against its success. The history that it left, however, illuminates how aerospace science, cultural politics, and gender relations intersected in the early 1960s.

CONCLUSION

Lovelace's Woman in Space Program existed at a time when the United States unknowingly stood poised on the brink of a revolution in gender consciousness. At this moment in the late 1950s and early 1960s, the debate over whether women could be astronauts presaged the debate about recognizing women's full participation in public life. At the same time, the story of selected female pilots' being thrust into the public eye revealed a glimpse of active women's lives before women's liberation. As a result, the Woman in Space Program offers a new understanding of the history of women, science, culture, and politics at a key juncture in the cold war and in the twentieth century.

The Lovelace Woman in Space program existed (albeit briefly) in a tenuous historical space created when space research's relative flexibility combined with a budding awareness of women's issues as politically viable. During the space agency's first years, the direction and form of U.S. space exploration was widely contested. Scientists wanted methodical experimentation to take precedence over human spaceflight, while space boosters heralded the imminent colonizing of space and the beginning of a space age. Politicians and public figures either urged a crash program to match Soviet advances or counseled caution to avoid wasting money. Before President John F. Kennedy dedicated NASA to achieving a lunar landing, human spaceflight did not belong exclusively to the official space agency. But by mid-1961, as the U.S. space program became streamlined to put a man on the moon, the form and purpose of American space exploration solidified. Many non-NASA space projects faded. At the same time, the question of female astronauts caught the public's interest because it emerged on the eve of a breakthrough in thinking about women's public roles.

The unspoken shift can be seen in the critique that never came. Notably, throughout the three-year debate over female astronauts, those who opposed the cause never voiced the most obvious rebuttal: astronauts cannot be women. Public figures recognized that women's concerns could not be dismissed outright. Increasingly in the early 1960s, women's issues got a public airing even through their champions lacked the widespread political support needed to achieve real changes.

Such harbingers of the women's movement have been called "prefeminist agitation." During the same years when Dr. Lovelace invited pilots to participate in his Woman in Space Program, Betty Friedan interviewed her Smith College classmates about their persistent dissatisfaction with their status—a dilemma she called "the problem that has no name." Friedan's call to raised consciousness first appeared in print in 1963, the same year that Valentina Tereshkova struck her blow for gender equality.[1] Like the "protofeminism" evinced by otherworldly female characters in early 1960s television shows and the well-traced political roots from which the modern women's movement was organized,[2] Lovelace's Woman in Space Program exemplified a liminal moment, a time when crucial elements of later historical movements first became perceptible.

The key to tracing this transition may lie in a greater historical appreciation for the many women who engaged in active public work between World War II and the women's movement. For instance, the story of women and space presents an intriguing parallel to the well-documented history of women scientists. (One proponent of female researchers, Penn State psychologist Donald Ford, summarized the problem in the sciences at an April 1962 conference by calling for a "female John Glenn.") During the early 1960s, women who answered the call to become scientists and engineers in the service of the cold war found institutional barriers still frustrating their success. Yet many persevered.[3] But by the time the modern women's movement of the late 1960s created powerful advocates, a vocal constituency, and widespread recognition for feminist issues, the question of female astronauts had long since left the public consciousness.[4]

The history of Lovelace's Woman in Space Program also represents a little-noticed turning point in medical ideas about women's bodies. Because Lovelace's program considered women as capable in their own right rather than as flawed men, the Albuquerque tests revealed that women could exhibit high levels of physical fitness, emotional stability, and mental endurance. Despite the discouraging conclusions published in 1964, Lovelace's female astronaut tests marked one of the earliest systematic investigations of women's bodies as strong, competent, and efficient. As such, the Woman in Space Program stands as part of a multifaceted transformation in thinking about women's physical capabilities over the course of the twentieth century.

Such rethinking had wide-ranging implications for health care, schooling, military service, and sports. From the early 1960s onward, women led the way in challenging the received wisdom about their bodies. Throughout the women's movement, feminists reclaimed their physical powers by reexamining everything from beauty norms to reproductive health care and childbirth practices. The Boston Women's Health Book Collective compiled research to produce the indispensable reference and guidebook for women's health: *Our Bodies, Ourselves*. Gradually, new conceptions of women's bodies ended the era when medical experts excluded women from medical or pharmaceutical studies—and then asked them to accept those findings as guides to their own health without any adjustment.[5]

New understanding of female bodies also encouraged women and girls to exercise their abilities. In the second half of the twentieth century, women's participation in athletics soared. In 1972 Title IX of the Educational Amendments entitled female athletes to equal support in scholastic sports by mandating proportionate funding for high school and college athletics. A generation of women came to expect opportunities to play. Beginning with golf in 1950 and tennis in 1971, professional associations allowed female athletes to make a living from their physical abilities. By the turn of the twenty-first century, basketball and soccer joined the growing list of professional women's leagues. Female sports stars also gained access to lucrative endorsement contracts, although not at the same level as similarly recognized male athletes. (Venus Williams's 2001 Reebok deal for $40 million set a record for a female athlete but paled in comparison with many men's contracts.) All of this progress followed from fundamental reevaluations of women's physical capabilities over the course of the twentieth century.[6]

Unfortunately, women's physical abilities had to be rediscovered over and over. Just as Lovelace's Woman in Space Program failed to utilize the medical data collected from women serving in military or military-auxiliary forces during World War II, the findings collected by the Lovelace Foundation disappeared from use following the dismal publication of the project's findings. When NASA officials decided to consider admitting female astronaut candidates in the 1970s, the agency started from scratch. Without using the Lovelace results, it rediscovered women's fitness for spaceflight by conducting bed rest studies on volunteer nurses.[7]

NASA finally recruited female astronauts because by the late 1970s it needed to sell its missions in a transformed political environment. In the years following the space agency's founding in the late 1950s, the social and political upheaval of the 1960s and early 1970s wrought massive changes in American society. Legal protections against workplace discrimination forced NASA to diversify its hiring in order to comply with federal statutes. At the same time, an ongoing war in Vietnam combined with a contracting economy to place government dollars at a premium. Although money had flowed when NASA dedicated itself to answering a fallen president's call for a moon landing, as the Apollo program ended the agency competed for dwindling funds.

When the space agency began choosing its first class of astronauts for the new space shuttle program in 1978, the all-white, all-male, clean-cut, military-style pilots who embodied cold war strength in the early 1960s seemed hopelessly out of touch with the liberated, racially diverse world of the late 1970s. In addition to the political changes that arose from student and urban uprisings, the identity politics advanced by the continued movements for civil rights, Black Power, women's liberation, Chicano recognition, and American Indian rights (to name just a few) left indelible marks on the nation's expectations. In order to appeal to the public for much-needed political and financial support, NASA wanted astronauts who reflected the face of the nation.[8]

In the end, the space agency needed to actively recruit such diversity. Simply opening the application process did not attract a sufficient number of female and minority-identified candidates because the previous history of exclusion had left a legacy of mistrust. To promote the agency's new slogan of "space for everyone," NASA hired Nichelle Nichols, the African American actress who played *Star Trek*'s Lieutenant Uhura. For several months in 1977, she promoted the agency's new inclusiveness in a nationwide publicity tour. To promote NASA's new image as a welcoming agency, Nichols tapped into science fiction's vision of racial and gender diversity in outer space.

The need for such advertising demonstrated the crucial importance of widespread social and political support. Even with legal protections in place—and with support for racial and gender diversity coming both from within the agency and from key space policy makers—the inclusion of women and minorities as astronauts still required that NASA invite them to participate. The contrast between the late 1970s and the early 1960s highlights just how much the historical context influenced space policy decisions. Without sustained advocacy and widespread support, the Woman in Space Program ended abruptly, never to be resumed.

Despite the attention to Sally Ride as the first American woman in space in 1983, the truest successors of the women pilots that Lovelace tested came in the selection of the first female pilot astronauts. To handle the varying tasks required to optimize shuttle missions, NASA created two tracks for astronauts: the mission specialists like Ride, who held PhDs and MDs and conducted experiments, and the pilot astronauts, who commanded and piloted the missions. The latter group of astronaut candidates rose through the ranks of military flying, the very training grounds that remained closed to women from the 1944 WASP disbandment until the early 1970s.

From her earliest missions as a shuttle pilot, Lt. Col. Eileen Collins allowed the women who participated in Lovelace's Woman in Space Program to reconnect with their history. With the help of her family, Collins worked hard to include these special women in her achievements. For those who were denied opportunities almost forty years earlier because they were the wrong sex, witnessing a fellow woman pilot going into space made all the difference.

To celebrate Eileen Collins's 1999 flight as the first woman commander of a space shuttle, NASA officials dedicated the launch of STS-93 to female aviation pioneers. The space agency contacted the Women Airforce Service Pilots (WASP), inviting some of its members to watch the launch in person at the Kennedy Space Center. Other famous women fliers also received invitations to NASA's Florida launch site. Betty Skelton Frankman, the aerobatic flying champion who graced *Look* magazine's cover as a possible answer to the question, "Should a Girl Be First in Space?" made the trip. For female pilots who knew each other in the 1950s from air races or women's flying organizations such as the Ninety-Nines and the Whirly-Girls, the gathering rekindled old friendships. For NASA, the tribute to pioneering women pilots provided a fitting historical backdrop for the space agency's first female commander.

As she had done before, Collins invited the women from Lovelace's Woman in Space Program to attend her July 1999 launch. For the first time, however, women who had taken the Lovelace tests but had not made the list of Pensacola candidates joined the rest of the Lovelace women on NASA's invitation list. Fran Bera and Georgiana T. McConnell met with Jerrie Cobb, Wally Funk, Jane Hart, Sarah Gorelick Ratley, Jerri Sloan Truhill, and Rhea Hurrle Woltman at Cape Canaveral for the anticipated launch date. The women received special VIP passes entitling them to tours of the Kennedy Space Center and a place on the observation grandstands during the launch. They received a NASA briefing on the shuttle's mission. In special recognition of their trailblazing roles, the women from Lovelace's Woman in Space Program also attended a private reception with Eileen Collins's family before the launch.

Delays foiled several of the women's plans to watch Collins's shuttle take off. On the announced launch date, the countdown reached four seconds when a sensor aborted the launch sequence. The shuttle would not launch until NASA engineers determined whether the alert indicated a serious problem. On a second attempt, a day later, the women waited as towering thunderheads blew toward the launch site. When it became apparent that the storm would not pass quickly enough, NASA officials postponed the scheduled flight a second time. After these delays, several of the women had to leave Florida.

For those who remained to see it in person, however, the launch of STS-93 on July 22, 1999, surpassed all their expectations. Georgiana McConnell wrote of her experience: "The third try was great. It was exciting beyond belief. We were standing on the ground and you could feel it all over your body as the fire took off! . . . It brought tears to your eyes." For these special guests of the space agency, the visceral power of the shuttle's massive engines matched the emotional thrill of seeing the first woman command an American spacecraft. It also provided an appropriate epilogue for Lovelace's Woman in Space Program. Pioneering female fliers joined the women pilots who participated in America's first women's space fitness tests to witness the crowning achievement of the United States' first woman *pilot* astronaut.[9]

Their feelings were similar to the very first time the women of Lovelace's Woman in Space Program watched Collins rocket into space. As Jerri Truhill recalled in 1995, "It was just about the most moving thing, I believe, I've ever experienced. It—she took something from each of us into space, except me, and I told her that she was carrying my dreams, that was all that was necessary."[10]

NOTES

Abbreviations

AAAS	"Developing U.S. Launch Capability: The Role of Civil-Military Cooperation," American Association for the Advancement of Science conference, Washington, DC, 5 November 1999
ITRI	Inhalation Toxicology Research Institute, Kirtland Air Force Base, Albuquerque, New Mexico
JAeS	*Journal of the Aeronautical Sciences*
JCobbP	Jerrie Cobb Papers, Ninety-Nines International Organization of Women Pilots Headquarters, Will Rogers Airport, Oklahoma City, Oklahoma
JCP	Jacqueline Cochran Papers, Dwight D. Eisenhower Presidential Library, Abilene, Kansas
JFK	John F. Kennedy Presidential Library, Boston, Massachusetts
LBJ	Lyndon Baines Johnson Presidential Library, Austin, Texas
NASA	Historical Collection, National Aeronautics and Space Administration Headquarters History Office, Washington, DC
NASM	Aeronautics Archives, National Air and Space Museum, Smithsonian Institution, Washington, DC
TWU	Women Airforce Service Pilots archives in Woman's Collection, Texas Woman's University, Denton, Texas
UNM	New Mexico History of Medicine Project, Medical Center Library, University of New Mexico, Albuquerque, New Mexico

Introduction

1. "From Aviatrix to Astronautrix," *Time*, August 29, 1960, 41; photograph in Jerrie Cobb file, Historical Collection, National Aeronautics and Space Administration Headquarters History Office, Washington, DC.

2. Although Cobb used "FLATs" (fellow lady astronaut trainees) in her correspondence, that acronym—sometimes adjusted to "first lady astronaut trainees"—emphasizes a group identity that did not exist until thirty years after the fact. By the time the successful participants met in the 1990s, two of the women had died. As a result, referring to the women collectively presents a problem. The newly minted "Mercury Thirteen" inaccurately suggests an association with NASA's Project Mercury. The most accurate designation remains the name that Lovelace gave the overall testing initiative: the Woman in Space Program.

3. Elaine Tyler May, *Homeward Bound: American Families in the Cold War Era* (New York: Basic Books, 1988); Joanna Meyerowitz, ed., *Not June Cleaver: Women and Gender in Postwar America, 1945–1960*, Critical Perspectives on the Past (Philadelphia: Temple University Press, 1994); Wini Breines, *Young, White, and Miserable: Growing Up Female in the Fifties* (Chicago: University of Chicago Press, 2001); Anna Greenwood Creadick, "Keeping Up Appearances: 'Normality' in Postwar United States Culture, 1945–1963" (PhD diss., University of Massachusetts, Amherst, 2002).

4. May, *Homeward Bound*, 16–19.

5. A small sampling of supporting scholarship includes May, *Homeward Bound*; Stephen Whitfield, *The Culture of the Cold War* (Baltimore: Johns Hopkins University Press, 1990); Meyerowitz, *Not June Cleaver*; and Robert D. Dean, *Imperial Brotherhood: Gender and the Making of Cold War Foreign Policy* (Amherst: University of Massachusetts Press, 2001).

6. Ruth Rosen, *The World Split Open: How the Women's Movement Changed America* (New York: Penguin Books, 2000).

7. Susan J. Douglas, *Where the Girls Are: Growing Up Female with the Mass Media* (New York: Time Books, 1994), 126–27. See also Margaret A. Weitekamp, "The 'Astronautrix' and 'The Magnificent Male': Jerrie Cobb's Quest to Be the First Woman in America's Manned Space Program," in *Impossible to Hold*, ed. Lauri Umanski and Avital Bloch (New York: New York University Press, 2004).

Chapter 1. Randy Lovelace and Jackie Cochran

1. Despite carrying no monetary award, the Collier Trophy's illustrious recipients and its presentation by the sitting U.S. president built its prestige. Pamela E. Mack, ed., *From Engineering Science to Big Science: The NACA and NASA Collier Trophy Research Project Winners*, NASA History Series (Washington, DC: National Aeronautics and Space Administration, 1998), xiii, xv; Jacqueline Cochran to W. R. Lovelace II, November 18, 1940, Jacqueline Cochran Papers, Dwight D. Eisenhower Presidential Library, Abilene, Kansas (hereafter cited as JCP). Lovelace to Cochran, November 25, 1940; Cochran to Lovelace, December 5, 1940; Walter M. Boothby to Cochran, De-

cember 5, 1940; Cochran to Lovelace, December 9, 1940; and Lovelace to Cochran, December 11, 1940, all JCP.

2. The gendered symbolism represented "the genius of Man (chief figure), [which] having conquered Gravity (male figure) and Contrary Wings (female figure) and having touched the bird and found its secrets, soars from the earth a conqueror." Frederick R. Neeley, "The Robert J. Collier Trophy," in *For the Greatest Achievement: A History of the Aero Club of America and the National Aeronautic Association*, ed. William Robie (Washington, DC: Smithsonian Institution Press, 1993), 83.

3. Photograph reproduced in Jake W. Spidle Jr., *The Lovelace Medical Center: Pioneer in American Health Care* (Albuquerque: University of New Mexico Press, 1987), 56. See also "President Presents Annual Aviation Award," *New York Times*, December 18, 1940.

4. Photograph reproduced in Richard G. Elliott, "'On a Comet Always': A Biography of Dr. W. Randolph Lovelace II," *New Mexico Quarterly* 36 (1966–67): n.p.

5. Spidle, *Lovelace Medical Center*, 4–5, 8, 16, 50; Jake W. Spidle Jr., *Doctors of Medicine in New Mexico* (Albuquerque: University of New Mexico Press, 1986), 34; Elliott, "'On a Comet,'" 356.

6. Colleagues called William Randolph Lovelace I and William Randolph Lovelace II either "Dr. L-I" and "Dr. L-II" or "Uncle Doc" and "Randy." Elliott, "'On a Comet,'" 358–59, 371; Spidle, *Lovelace Medical Center*, 31, 52.

7. Elliott, "'On a Comet,'" 360.

8. Adrianne Noe, "Medical Principle and Aeronautical Practice: American Aviation Medicine to World War II" (PhD diss., University of Delaware, 1989), 2; Elliott, "'On a Comet,'" 384.

9. Walter M. Boothby, W. Randolph Lovelace II, and Otis O. Benson, "High Altitude and Its Effect on the Human Body, II," *Journal of the Aeronautical Sciences* 7 (October 1940): 525 (hereafter the *Journal of the Aeronautical Sciences* will be cited as *JAeS*). "Editorial: Recent Developments in Use and Administration of Oxygen in Aviation and Therapeutics," *Annals of Internal Medicine* 12 (October 1938): 560–61, JCP.

10. P. R. Bassett, "Passenger Comfort in Air Transportation," *JAeS* 2 (March 1935): 48. See also Michael E. Gluhareff, "High Altitude Problems," *JAeS* 3 (March 1936): 154–55.

11. Elliott, "'On a Comet,'" 361.

12. Charles A. Dempsey, *Fifty Years of Research on Man in Flight* (Wright-Patterson Air Force Base, OH: Aerospace Medical Research Laboratory, 1985), 1; Elliott, "'On a Comet,'" 362; "Collier Prize Goes to Three Doctors: Army Physician and Two Mayo Clinic Aides, as Well as Air Lines Share Reward; Effect on Blood Is Aim; The 'Blacking Out' of Fliers in Stratosphere Flights Is Remedied by Mask," *New York Times*, November 15, 1940.

13. Walter M. Boothby, W. Randolph Lovelace II, and Otis O. Benson, "High Altitude and Its Effect on the Human Body, I," *JAeS* 7 (September 1940): 461–68; Boothby, Lovelace, and Benson, "High Altitude and Its Effect, II," 524–30; J. W. Heim, "Physiologic Considerations Governing High Altitude Flight," *JAeS* 5 (March 1938): 190; Harry

G. Armstrong, "The Influence of Aviation Medicine on Aircraft Design and Operation," *JAeS* 2 (March 1938): 197.

14. The Reminiscences of Doctor William R. Lovelace II, interview with Kenneth Leish, July 1960, Aviation Project, Columbia University Oral History Collection, Columbia University, New York City, 4; Donald E. Kilgore, interview with author, edited transcribed tape recording, Albuquerque, New Mexico, April 30, 1997; Boothby, Lovelace, and Benson, "High Altitude and Its Effects, I," 466. See also Ross A. McFarland, *Human Factors in Air Transportation: Occupational Health and Safety* (New York: McGraw-Hill, 1953), x–xi.

15. W. Randolph Lovelace II to Jacqueline Cochran, September 10, 1940, JCP.

16. Jacqueline Cochran, *The Stars at Noon* (Boston: Little, Brown, 1954), 3. Cochran was probably born about 1906. See Jacqueline Cochran and Maryann Bucknum Brinley, *Jackie Cochran: An Autobiography* (New York: Bantam Books, 1987).

17. Cochran and Bucknum Brinley, *Jackie Cochran*, 49. *The Stars at Noon* omitted the account of Jackie's adopted surname.

18. Cochran, *Stars at Noon*, 7; Cochran and Bucknum Brinley, *Jackie Cochran*, 344.

19. Cochran, *Stars at Noon*, 29.

20. *Current Biography* (New York: H. W. Wilson, 1941), s.v. "Odlum, Floyd B(ostwick)," 62; Cochran, *Stars at Noon*, 40, 42–43.

21. Cochran, *Stars at Noon*, 44, 45.

22. Claudia M. Oakes, *United States Women in Aviation, 1930–1939* (Washington, DC: Smithsonian Institution Press, 1991), 30–31.

23. Cochran, *Stars at Noon*, 49.

24. Cochran, *Stars at Noon*, 62. Louise Thaden credited the Bendix organizers' admission of women to Clifford Henderson's advocacy on their behalf. Oakes, *United States Women in Aviation*, 30–31.

25. Oakes, *United States Women in Aviation*, 30, 37; Dean Jaros, *Heroes without Legacy: American Airwomen, 1912–1944* (Niwot: University of Colorado Press, 1993), 51; Cochran, *Stars at Noon*, 66; *Current Biography* (New York: H. W. Wilson, 1940), s.v. "Cochran, Jacqueline," 183.

26. Cochran, *Stars at Noon*, 51, 53, 55; Jaros, *Heroes without Legacy*, 50, 51. "The Gee Bee racers, built by the Granville Brothers of Springfield, Massachusetts, were the quintessential racing aircraft of the 1930s. Their sometimes deadly speed was due to their short, bulky fuselages, clipped wings, and high-horsepower engines." Oakes, *United States Women in Aviation*, 31.

27. Odlum and Cochran also experimented with "automatic writing," guided writing while in a trancelike state. Cochran, *Stars at Noon*, 86–87, 90.

28. Susan Ware, *Still Missing: Amelia Earhart and the Search for Modern Feminism* (New York: W. W. Norton, 1993), 24, 122, 125, 169, 145, 148, 159–60.

29. Ware, *Still Missing*, 50–51.

30. Donald E. Kilgore, interview with author, April 30, 1997; Jane Hart, interview with author, edited transcribed tape recording, Mackinac City, Michigan, October 7, 1997; Cochran, *Stars at Noon*, 14–15.

31. Cochran and Bucknum Brinley, *Jackie Cochran*, 121; Cochran, *Stars at Noon*, 66.

32. Kathy Peiss, *Hope in a Jar: The Making of America's Beauty Culture* (New York: Metropolitan Books, 1998), 186, 187; Cochran, *Stars at Noon*, 66.

33. Cochran and Bucknum Brinley, *Jackie Cochran*, 20; *Current Biography* (1940), s.v. "Cochran, Jacqueline," 182.

34. Cochran, *Stars at Noon*, 61–62.

35. Elliott, "'On a Comet,'" 361.

36. Robie, *For the Greatest Achievement*, 83; G. deFreest Larner to Cochran, June 7, 1940, JCP, 2.

37. W. A. Patterson to George Lewis, July 31, 1940; S. A. Stewart to Cochran, August 7, 1940; and "Nominations for the 1939 Collier Trophy Award," n.d., all JCP.

38. Cochran, Open letter to Collier Trophy Award Committee, August 12, 1940, JCP.

39. As assistant chief of the Air Services, Gen. Billy Mitchell accused army and navy officials of stifling military aviation because they feared air power would displace sea power. Even after being demoted, Mitchell criticized his superiors publicly. Although Mitchell's court-martial made him a popular hero, he was forced to resign his commission. William Mitchell, *Winged Defense: The Development and Possibilities of Modern Air Power—Economic and Military* (New York: G. P. Putnam's Sons, 1925). The episode inspired a 1955 feature film starring Gary Cooper as Mitchell. Otto Preminger, dir., *The Court-Martial of Billy Mitchell* (Warner Brothers, 1955). Cochran, August 12, 1940, JCP; Cochran to Lewis, August 1, 1940, JCP, 2.

40. Cochran to Lewis, August 1, 1940, JCP, 1; Cochran, August 12, 1940, JCP, 3.

41. Cochran, August 12, 1940, JCP, 2. See also Cochran to William B. Wheatley, September 16, 1940; Cochran to E. V. Rickenbacker, September 16, 1940; and Cochran to Rickenbacker, September 19, 1940, all JCP.

42. Cochran to Rube Fleet, September 16, 1940; Cochran to Frank Fuller, September 16, 1940; and Cochran to William B. Wheatley, September 16, 1940, all JCP.

43. Cochran to Fleet, September 16, 1940; Cochran to William B. Wheatley, September 16, 1940; and Frank Fuller Jr. to Cochran, September 11, 1940, all JCP.

44. Telegram, Lovelace to Cochran, August 10, 1940, JCP; G. de Freest Larner to Cochran, August 19, 1940, JCP.

45. Walter M. Boothby to Odlum, August 21, 1940, JCP. See also Lovelace to Odlum, August 21, 1940, JCP.

46. Cochran to Boothby, August 29, 1940, JCP.

47. Cochran to Lovelace, September 11, 1940, JCP, 1; Cochran to William R. Enyart, September 11, 1940, JCP, 2–3.

48. G. de Freest Larner to Cochran, September 18, 1940, JCP; Cochran to Lovelace, September 11, 1940, JCP; Harry G. Armstrong, *The Principles and Practice of Aviation Medicine* (Baltimore: Williams and Wilkins, 1939).

49. Harry Bruno to Cochran, August 12, 1940; Cochran to Capt. Harry G. Armstrong, September 18, 1940; H. H. Arnold to Cochran, September 24, 1940; and Cochran to Rickenbacker, September 16, 1940, all JCP.

50. "Collier Trophy Nominations to Be Considered at September 26, 1940 meeting

of Collier Trophy Committee," n.d., JCP; Denver Lindley, "Take a Deep Breath," *Collier's*, November 23, 1940, 16–17, 67; "Collier Prize," *New York Times*, November 15, 1940.

51. Telegram, Boothby and Lovelace to Cochran, November 15, 1940, 6:27 p.m.; telegram, Boothby and Lovelace to Cochran, November 15, 1940, 11:19 p.m.; and Lovelace to Cochran, November 15, 1940, all JCP.

52. Frederick Grahman, "Air Currents," *New York Times*, November 17, 1940, sect. 10; Lindley, "Deep Breath."

Chapter 2. Aviation and Aerospace Medicine

1. Richard G. Elliott, "'On a Comet Always': A Biography of W. Randolph Lovelace II," *New Mexico Quarterly* 36 (1966–67): 363; "Man Who Made 40,200-Foot Parachute Jump Visits Here," *Houston Chronicle*, n.d., Jacqueline Cochran Papers, Dwight D. Eisenhower Presidential Library, Abilene, Kansas (hereafter cited as JCP).

2. "Man Who Made Parachute Jump Visits."

3. Adrianne Noe, "Medical Principle and Aeronautical Practice: American Aviation Medicine to World War II" (PhD diss., University of Delaware, 1989).

4. The Reminiscences of Doctor William R. Lovelace II, interview with Kenneth Leish, July 1960, Aviation Project, Columbia University Oral History Collection, Columbia University, New York City, 10.

5. "Army Doctor's Record Parachute Jump," *Life*, August 9, 1943, 69; Elliott, "'On a Comet,'" photographs between pages 360 and 361; Martin W. Bowman, *The USAAF Handbook, 1939–1945* (Mechanicsburg, PA: Stackpole Books, 1997), 171, 173–74. See also Robert J. Benford, *The Heritage of Aviation Medicine: An Annotated Directory of Early Artifacts* (Washington, DC: Aerospace Medical Association, 1979), 25.

6. Robert Joseph Benford, *Doctors in the Sky: The Story of the Aero Medical Association* (Springfield, IL: Charles Thomas, 1955), 154.

7. "Army Flier Jumps from 40,200 Feet; He Proves Oxygen Equipment Will Work Satisfactorily at Very High Altitudes; His Left Hand Is Frozen; Jerk of 'Chute Flips off Glove at 50 Below as He Plunges to Set Up a New Record," *New York Times*, July 1, 1943.

8. Lovelace Reminiscences, 10–11, 12; Jake Spidle Jr., *The Lovelace Medical Center: Pioneer in American Health Care* (Albuquerque: University of New Mexico Press, 1987), 59.

9. Benford, *Doctors in the Sky*, 156.

10. Charles A. Dempsey, *Air Force Aerospace Medical Research Laboratory: 50 Years of Research on Man in Flight* (Wright Patterson Air Force Base, OH: Aerospace Medical Research Laboratory, 1985), xxv.

11. Noe, "Medical Principle and Aeronautical Practice," 201, 215; Lovelace Reminiscences, 8, 14.

12. Elliott, "'On a Comet,'" 363.

13. Spidle, *Lovelace Medical Center*, 65; Elliott, "'On a Comet,'" 367.

14. "Mayo-like Center Founded: New Mexico Foundation Will Have as Its Major Interest Expansion of the Lovelace Clinic's Cancer Service and Research in Aviation Medicine," *Science News Letter*, October 4, 1947, 210.

15. "Lovelace Clinic in New Mexico Gets $1,000,000 for Research: Foundation to Be Operated at the Outset from Earnings of Doctors in Excess of Expenses—Odlum Is Head of Board," *New York Times*, September 26, 1947; Spidle, *Lovelace Medical Center*, 75–79.

16. Spidle, *Lovelace Medical Center*, 106, 107.

17. Shirley Thomas, "Hubertus Strughold: The Father of Space Medicine Whose Dramatic Advanced Planning Encompasses the Universe," in *Men of Space: Profiles of the Leaders in Space Research, Development, and Exploration*, vol. 4 (Philadelphia: Chilton, 1962), 251; Lloyd S. Swenson, James M. Grimwood, and Charles C. Alexander, *This New Ocean: A History of Project Mercury*, NASA SP-4201 (Washington, DC: Government Printing Office, 1966), 34; Otis O. Benson Jr., "Preface," in *Physics and Medicine of the Atmosphere and Space*, ed. Otis O. Benson Jr. and Hubertus Strughold (New York: John Wiley, 1960), ix.

18. Insertion in original. A. H. Schwichtenberg, oral history interview conducted by Jake Spidle, February 20, 1985, New Mexico History of Medicine Project, Medical Center Library, University of New Mexico, Albuquerque, New Mexico, 4 (hereafter cited as UNM). Coincidentally, Wilson's last day in office occurred just five days after *Sputnik*'s launch. Robert A. Divine, *The Sputnik Challenge: Eisenhower's Response to the Soviet Satellite* (New York: Oxford University Press, 1993), 20.

19. Schwichtenberg oral history interview, 4; Gen. Donald Flickinger, interview by John Pitts, October 18, 1979, transcript, Historical Collection, National Aeronautics and Space Administration Headquarters History Office, Washington, DC, 9 (hereafter cited as NASA).

20. Harry G. Armstrong, "Foreword," in *Physics and Medicine of the Upper Atmosphere: A Study of the Aeropause*, ed. Clayton S. White and Otis O. Benson Jr. (Albuquerque: University of New Mexico Press, 1952), xiv.

21. Thomas, "Hubertus Strughold," 233; Hubertus Strughold, "Basic Environmental Problems relating Man and the Highest Regions of the Atmosphere as Seen by the Biologist," in White and Benson, *Physics and Medicine*, 31, 35 (graph).

22. Strughold, "Environmental Problems," 31, 35; italics in the original; Clayton S. White, "Introduction," in White and Benson, *Physics and Medicine*, 3–4. Although "aeropause" was used differently before 1951, Strughold's interpretation became the accepted scientific definition.

23. Wayne Lee, *To Rise from Earth: An Easy to Understand Guide to Spaceflight* (New York: Facts on File, 1995), 11–12. Present-day scientists define space as beginning at 400,000 feet or 75.76 miles above the earth's surface.

24. James J. Haggerty, "Fastest Man on Earth," *Collier's*, June 25, 1954, 28–29; "Fantastic Catch in the Sky, Record Leap towards Earth: Space Race Soars with a Vengeance," *Life*, August 29, 1960, 20–25.

25. Kilgore interview, 1, 2.

26. Ibid., 2.

27. Dr. Ulrich Luft, oral history interview conducted by Jake Spidle, October 11 and 16, 1985, UNM, 12; Dr. Robert Secrest, oral history interview conducted by Jake Spidle, July 8, 1996, UNM, 10.

28. Spidle, *Lovelace Medical Center*, 111, 113. The AEC contract and subsequent AEC work continued for thirty-five years. The Inhalation Toxicology Laboratory remains at Kirtland Air Force Base as one of the last vestiges of the Lovelace Foundation. Michael R. Beschloss, *Mayday: Eisenhower, Khrushchev, and the U-2 Affair* (New York: Harper and Row, 1986), 109.

29. Gen. Bernard Schriever, "Reflections," in "Developing U.S. Launch Capability: The Role of Civil-Military Cooperation," American Association for the Advancement of Science conference, Washington, DC, November 5, 1999 (hereafter cited as AAAS).

30. Dr. Simon Ramo, "Reflections," AAAS. Ramo is the *R* in TRW, the aerospace firm.

31. Shirley Thomas, "Don D. Flickinger: With Zest and Dedication, This Energetic Doctor Has Long Concentrated on the Problems of Man's Survival in the Hostile Environment of Space," in *Men of Space: Profiles of the Leaders in Space Research, Development, and Exploration*, vol. 3 (Philadelphia: Chilton, 1961), 77; Roger D. Launius, *NASA: A History of the U.S. Civil Space Program* (Malabar, FL: Krieger, 1994), 38.

32. Included in Democratic National Committee, Cartoon Clippings File, Prepresidential Papers, John F. Kennedy Presidential Library, Boston, Massachusetts (hereafter cited as JFK).

33. Divine, *Sputnik Challenge*; Launius, *NASA: A History*, 29–32.

34. The air force maintained a human spaceflight program throughout most of the 1960s, first with Dyna-soar, then with its 1963 replacement the Manned Orbiting Laboratory (MOL). When the air force canceled the project in 1969, NASA incorporated several MOL astronauts into its astronaut corps.

35. Roger Bilstein, *Orders of Magnitude: A History of the NACA and NASA, 1915–1990*, NASA History Series (Washington, DC: National Aeronautics and Space Administration Office of Management, 1989), 1.

36. Launius, *NASA: A History*, 32–34.

37. Flickinger interview, October 18, 1979, 9. See also Mae Mills Link, *Space Medicine in Project Mercury* (Washington, DC: National Aeronautics and Space Administration, Scientific and Technical Information Division, 1965), 39–40; Hugh Dryden, Deputy Administrator, to James E. Webb, Administrator, memorandum, April 30, 1962, Vice-Presidential Papers, Lyndon Baines Johnson Presidential Library, Austin, Texas (hereafter cited as LBJ).

38. Launius, *NASA: A History*, 38.

39. Kilgore interview.

40. Tom Wolfe, *The Right Stuff* (New York: Farrar, Straus, Giroux, 1979). See also Randy Lovelace, "Duckings, Probings, Checks That Proved Fliers' Fitness," *Life*, April 20, 1959.

41. Joseph D. Atkinson Jr. and Jay M. Shafritz, *The Real Stuff: A History of NASA's Astronaut Recruitment Program* (New York: Praeger, 1985), 10.

42. Atkinson and Shafritz, *Real Stuff*, 2; Swenson, Grimwood, and Alexander, *This New Ocean*, 160.

43. Atkinson and Shafritz, *Real Stuff*, 18.

44. Spidle, *Lovelace Medical Center*, 137; Atkinson and Shafritz, *Real Stuff*, 43; A. H. Schwichtenberg, Donald D. Flickinger, and W. Randolph Lovelace II, "Medical Machine Record Cards, Their Development and Use in the Astronaut Selection Program," original office copy, Inhalation Toxicology Research Institute Library, Kirtland Air Force Base, Albuquerque, New Mexico. This article was published in the November 1959 *U.S. Armed Forces Medical Journal.*

45. John A. Pitts, *The Human Factor: Biomedicine in the Manned Space Program to 1980*, NASA History Series (Washington, DC: NASA Science and Technical Information Branch, 1985), 19.

46. Atkinson and Shafritz, *Real Stuff*, 44–45.

Chapter 3. Female Pilots and Postwar Women's Aviation

1. The Texas Woman's University WASP collection in Denton, Texas, holds Jean F. Hixson's papers and uniform. Molly Merryman, *Clipped Wings: The Rise and Fall of the Women Airforce Service Pilots (WASPs) of World War II* (New York: New York University Press, 1998), 103–6, 115.

2. Pauline Vincent (Jean Hixson's sister), oral history interview, edited transcribed tape recording, Denver, Colorado, March 14, 1998, 8.

3. Pauline Vincent interview, 3; Florence E. Curry, "Miss Jean Hixson Takes Part in Weightlessness Studies: All a Part of Aerospace Training," *Hoopeston (IL) Chronicle-Herald*, February 7, 1964, WASP archives in Woman's Collection, Texas Woman's University, Denton, Texas (hereafter cited as TWU).

4. Dominick A. Pisano, *To Fill the Sky with Pilots: The Civilian Pilot Training Program, 1939–46* (Urbana: University of Illinois Press, 1993), 56, 57; Deborah Douglas, *United States Women in Aviation, 1940–1985* (Washington, DC: Smithsonian Institution Press, 1991), 6.

5. Douglas, *Women in Aviation*, 6–7. Some women made and honored the pledge. Although she washed out, Jean Ross Howard later fulfilled her CPTP promise by joining the WASP. Shelley Davis, "Taking a Spin with the Whirly-Girls," *Aviation for Women*, September–October 2000, 33.

6. Olga Gruhzit-Hoyt, *They Also Served: American Women in World War II* (New York: Birch Lane Press, 1995), 61–62, 102, 141.

7. Douglas, *Women in Aviation*, 29–30.

8. Douglas, *Women in Aviation*, 43.

9. Jacqueline Cochran, *The Stars at Noon* (Boston: Little, Brown, 1954), 121; Merryman, *Clipped Wings*, 75, 110; Douglas, *Women in Aviation*, 55.

10. Ke-chin Wang to Jean Hixson, October 12, 1944, TWU.

11. Douglas, *Women in Aviation*, 103.

12. Ann Carl, *A WASP among Eagles: A Woman Military Test Pilot in World War II* (Washington, DC: Smithsonian Institution Press, 1999), 97, 113. Jacqueline Cochran and Maryann Bucknum Brinley, *Jackie Cochran: An Autobiography* (New York: Bantam Books, 1987), 274.

13. Commercial and military aviation reopened to women in 1973. Emily Howell

Warner became the first full-time pilot hired by a major airline (since Helen Richey resigned from Central Airlines in 1934) when Frontier Airlines hired her in January. Anticipating that the Equal Rights Amendment would pass, navy admiral Elmo Zumwalt urged the navy to open its flight training to women voluntarily. The air force followed soon after. In 1976 the army admitted women to its flight training. Henry Holden and Lori Griffith, *Ladybirds II: The Continuing Story of American Women in Aviation* (Mount Freedom, NJ: Blackhawk, 1993), 85, 256.

14. Jerri Sloan Truhill, oral history interview, edited transcribed tape recording, Richardson, Texas, September 27, 1997, 8.

15. Jane B. Hart, oral history interview, edited transcribed tape recording, Mackinaw City, Michigan, October 7, 1997; Jane B. Hart, questionnaire completed for author, June 4, 1999.

16. Sarah Gorelick Ratley, oral history interview, edited transcribed tape recording, Overland Park, Kansas, June 7, 1997, 14.

17. Virginia Holmes, telephone interview with author, December 17, 1997; Bernice Steadman, oral history interview, edited transcribed tape recording, Traverse City, Michigan, October 8, 1997, 4.

18. Gene Nora Stumbough Jessen, oral history interview, edited transcribed tape recording, Boise, Idaho, May 24, 1997, 1.

19. "Missions for America: The Civil Air Patrol Story," Civil Air Patrol recruiting pamphlet, ca. 1998, 1, 2.

20. Douglas, *Women in Aviation*, 9, 10, 74; "Missions for America," 2.

21. Irene Leverton, oral history interview, edited transcribed tape recording, Chino Valley, Arizona, November 22, 1997, 1–2.

22. Georgiana T. McConnell, questionnaire completed for author, December 2, 1997.

23. Jessen oral history interview, 2, 3, 5.

24. Douglas, *Women in Aviation*, 5; Marilyn C. Link, questionnaire completed for author, May 13, 1998.

25. "Biographical Information: Mary Wallace 'Wally' Funk, II," Aeronautics Archives, National Air and Space Museum, Smithsonian Institution, Washington, DC (hereafter cited as NASM).

26. Patricia Kelley Jetton, questionnaire completed for author, January 14, 1998.

27. Jessen oral history interview, 5; Ratley oral history interview, 5.

28. Davis, "Whirly-Girls," 34; Douglas, *Women in Aviation*, 76.

29. Davis, "Whirly-Girls," 35. Each new Whirly-Girl still receives a number.

30. Philip A. Hart to Andrew Hatcher, March 30, 1961, Executive Name Files, John F. Kennedy Presidential Library, Boston, Massachusetts (hereafter cited as JFK). See also Philip Hart to John F. Kennedy, May 15, 1961, Executive Name Files, JFK.

31. Rosanne Welch, *Encyclopedia of Women in Aviation and Space* (Santa Barbara, CA: ABC-CLIO, 1998), s.v. "Women's Air Derby," 250; Holden and Griffith, *Ladybirds II*, 31.

32. Douglas, *Women in Aviation*, 63; Hart oral history interview, 2.

33. Holden and Griffith, *Ladybirds II*, 279–80; Douglas, *Women in Aviation*, 64, 75.

34. "Female Astronaut: Aviatrix Looking Forward to Trips in Outer Space," *Dallas Times Herald*, October 29, 1961, clipping in Jerri Truhill's personal collection; Leverton oral history interview, 9.

35. Truhill oral history interview, 2.

36. Marilyn C. Link, questionnaire completed June 8, 1998, 1; Lu Curtis Hollander, ed., *History of the Ninety-Nines, Inc.* (Oklahoma City: Ninety-Nines, 1979).

37. Jerrie Cobb, *Jerrie Cobb: Solo Pilot* (Sun City Center, FL: Jerrie Cobb Foundation, 1997), 32–40.

38. Leverton oral history interview, 10–11, 4.

39. Edith Hills Coogler, "Comments on Space Trip. Myrtle Thompson Cagle: 'It Should Have Been Me,'" *Raleigh (NC) News and Observer*, July 9, 1963, Jacqueline Cochran Papers, Dwight D. Eisenhower Presidential Library, Abilene, Kansas (hereafter cited as JCP). Truhill oral history interview, 2; Steadman oral history interview.

40. Cobb, *Solo Pilot*, 77.

41. Eugenia Heise, "President's Column," *Ninety-Nines, Inc., International Organization of Women Pilots Newsletter*, October 1960, JCP.

42. Ratley oral history interview, 6; Georgiana T. McConnell, questionnaire, December 2, 1997. According to McConnell, "Upon return I planned to try to get a job in aviation. It didn't prove to be easy and I was unemployed for a number of months."

43. Eugenia Heise, "President's Column," *Ninety-Nines, Inc., International Organization of Women Pilots Newsletter*, September 1960, JCP, 1; Loma May, "The Lighter Side of the 99 European Tour," *Ninety-Nines, Inc., International Organization of Women Pilots Newsletter*, October 1960, JCP, 1; Ratley oral history interview, 6.

44. May, "Lighter Side of the 99 European Tour," 1.

45. Ratley oral history interview, 6.

Chapter 4. Betty Skelton, Ruth Nichols, and Jerrie Cobb

1. "Woman Space Pioneer," *Philadelphia Inquirer*, January 15, 1959, Aeronautic Archives, National Air and Space Museum, Smithsonian Institution, Washington, DC (hereafter cited as NASM). "Womanned Flight," *Christian Science Monitor*, February 13, 1959, NASM; "Women Best Men in Solo Space Tests," *Chicago Daily Tribune*, October 2, 1959, NASM; "Girl 'in Space' Six Days without a Hallucination," *Chicago Daily Tribune*, November 6, 1959, NASM; Cathryn Walters, Oscar A. Parsons, and Jay T. Shurley, "Differences in Male and Female Responses to Underwater Sensory Deprivation," undated draft of research results, Jerrie Cobb Papers, Ninety-Nines International Organization of Women Pilots, National Headquarters, Will Rogers Airport, Oklahoma City, Oklahoma (hereafter cited as JCobbP).

2. Lillian Levy, "Space Researcher Sees Female Solos Possible," *Washington Evening Star*, October 31, 1958, NASM.

3. "Little Spacegirls," *Time*, September 16, 1957, 69–70; "No Space for Powder Puffs," *Insider's Newsletter*, August 13, 1962, JCobbP.

4. Ben Kocivar, "The Lady Wants to Orbit," *Look*, February 2, 1960, NASM.

5. Kocivar, "Lady Wants to Orbit," 114, 116.

6. "Records—Betty Skelton Frankman," July 22, 1969, NASM; Henry M. Holden and Lori Griffith, *Ladybirds II: The Continuing Story of American Women in Aviation* (Mount Freedom, NJ: Black Hawk, 1993), 119–25.

7. Betty Skelton Frankman, interview by Carol L. Butler, July 19, 1999, Cocoa Beach, Florida, transcript, NASA Oral History Project, 12.

8. NASA food scientists still assert that flying more female astronauts would decrease overall payload weight. Charles Bourland, Johnson Space Center subsystem manager for space station food, NASA Headquarters, Washington, DC, April 30, 1998.

9. Kocivar, "Lady Wants to Orbit," 119.

10. Betty Skelton Frankman, interview, July 19, 1999, 11, 13.

11. "Recollections of Ruth Nichols," interview by Kenneth Leish, June 1960, New York Times Oral History program, Columbia University Oral History Collection, part 4 (1–219), Columbia University, New York City, 39; Dean Jaros, *Heroes without Legacy: American Airwomen, 1912–1944* (Niwot: University of Colorado Press, 1993), 32–34.

12. Nichols served as a reserve pilot for the unscheduled passenger airline. "Nichols, Ruth Rowland," brief biographical description, NASM, 1; Ruth Nichols, "'Behind the Ballyhoo': A Famous Pilot Explains the Showmen of the Air," *American Magazine*, March 1932, NASM, 43, 78, 80; "Recollections of Ruth Nichols," 27.

13. "Recollections of Ruth Nichols," 33, 34–35, 36–37. During the UNICEF flight, one of the airplanes ditched in the Irish Sea. Although a trawler rescued Nichols and others, seven of the fifty-two people on board died. Ruth Nichols, *Wings for Life* (Philadelphia: Lippincott, 1957); "Sets Record," *Post-Times Herald*, June 24, 1958, clipping, NASM.

14. "Recollections of Ruth Nichols," 41.

15. "Recollections of Ruth Nichols," 42–43.

16. The Air Rescue Service named its aerial rescue trophy the Don Flickinger Trophy. In 1941, Flickinger also served admirably as the medical officer on duty during the Japanese attack at Pearl Harbor. Shirley Thomas, "Don D. Flickinger: With Zest and Dedication, This Energetic Doctor Has Long Concentrated on the Problems of Man's Survival in the Hostile Environment of Space," in *Men of Space: Profiles of the Leaders in Space Research, Development, and Exploration*, vol. 3 (Philadelphia: Chilton, 1961), 83, 84–87; John A. S. Pitts, interview with Gen. Donald Flickinger, October 18, 1979, transcript, Historical Collection, National Aeronautics and Space Administration Headquarters History Office, Washington, DC, 3 (hereafter cited as NASA). Henry Fountain, "Dr. Donald D. Flickinger, 89, a Pioneer in Space Medicine," obituary, *New York Times*, 1997, undated clipping, NASA.

17. Jerrie Cobb and Jane Rieker, *Woman into Space: The Jerrie Cobb Story* (Englewood Cliffs, NJ: Prentice-Hall, 1963), 130, 132.

18. Jerrie Cobb, *Jerrie Cobb: Solo Pilot* (Sun City Center, FL: Jerrie Cobb Foundation, 1997), 46, 139; Cobb and Rieker, *Woman into Space*, 48; Martha Ackmann, *The Mercury 13: The Untold Story of Thirteen American Women and the Dream of Spaceflight* (New York: Random House, 2003), 6.

19. Jerrie Cobb, press release, "Project WISE," speech delivered at Space Symposium for Women, Air Force Association Convention, Las Vegas, Nevada, September 21, 1962, JCobbP; memorandum, Donald Flickinger to W. R. "Randy" Lovelace II, "Action Memorandum," December 20, 1959, JCobbP; Cobb and Rieker, *Woman into Space*.

20. Flickinger to Lovelace, December 20, 1959, JCobbP. The eight women Cobb selected were Fran Bera, Jane White, Geraldine Sloan, Marian Petty, Geraldyn Cobb, Marilyn Link, Barbara Erickson, and Betty Skelton.

21. Donald Flickinger to Jerrie Cobb, December 7, 1959, JCobbP.

22. "Ruth Rowland Nichols, Pioneer Flyer, Dies," *Washington Star*, September 26, 1960, NASM; Ackmann, *Mercury 13*, 70.

23. Memorandum, Donald Flickinger to W. R. "Randy" Lovelace II, "Action Memorandum," December 20, 1959, JCobbP; Kocivar, "Lady Wants to Orbit," 119.

24. Robert J. Benford, *The Heritage of Aviation Medicine: An Annotated Directory of Early Artifacts* (Washington, DC: Aerospace Medical Association, 1979), 29; Robert McG. Thomas Jr., "Russell Colley, the Designer of Spacesuits, Is Dead at 97," *New York Times*, February 8, 1996, NASA. Full and partial pressure suits served as precursors to space suits.

25. Flickinger to Lovelace, December 20, 1959, JCobbP, 2.

26. Ibid.

27. Cobb and Rieker, *Woman into Space*, 134. *Woman into Space* did not distinguish between Project WISE and the later Lovelace testing. Written for a popular audience by a reporter who interviewed Cobb's mother more than Cobb, the book missed that detail. Jerrie Cobb, oral history interview, edited transcribed tape recording, conducted at the Women in Aviation, International, conference in Denver, Colorado, March 13, 1998, 13.

28. "Woman Qualifies for Space Training," *Washington Post*, August 19, 1960, clipping, NASA; "A Woman Passes Tests Given to 7 Astronauts," *New York Times*, August 19, 1960.

29. "A Lady Proves She's Fit for Space Flight," *Life*, August 29, 1960, 72.

30. "Spacewoman Ready for Flights with Men," *Washington Star*, August 24, 1960, NASA; "From Aviatrix to Astronautrix," *Time*, August 29, 1960, 41.

31. "'Spacewoman' Would Let Men Come Along on Trip," *Baltimore Sun*, August 24, 1960, clipping compiled in *NASA Current News*, August 24, 1960, NASA; "Spacewoman Ready for Flights with Men," *Washington Star*, August 24, 1960, NASA; "Woman Predicts Coed Space Trips," untitled and undated New York paper, ca. 1959–60, NASA. See also Margaret A. Weitekamp, "The 'Astronautrix' and 'The Magnificent Male': Jerrie Cobb's Quest to Be the First Woman in America's Manned Space Program," in *Impossible to Hold: Women and Culture and the 1960s*, ed. Avital Bloch and Lauri Umansky (New York: New York University Press, 2004).

32. Cochran to Lovelace, November 28, 1960, Jacqueline Cochran Papers, Dwight D. Eisenhower Presidential Library, Abilene, Kansas (hereafter cited as JCP).

33. Cochran to Lovelace, December 27, 1960, JCP.

34. Cochran to Lovelace, November 28, 1960, JCP; Cochran to Lovelace, December 27, 1960, JCP.

35. Sarah Gorelick Ratley, oral history interview, edited transcribed tape recording, Overland Park, Kansas, June 7, 1997, 19–20. See also Gorelick to Cochran, July 20, 1961, JCP.

36. Cochran's age remained unclear, but she was over fifty by 1960. Jacqueline Cochran and Maryann Bucknum Brinley, *Jackie Cochran: An Autobiography* (New York: Bantam Books, 1987), 7; Jacqueline Cochran, *The Stars at Noon* (Boston: Little, Brown, 1954), 260–61.

37. Cochran to Eloise Engle, December 9, 1961, JCP, 3.

38. "Twin Pilots Describe Astronaut Tests, Tell Why They Desire Further Training," *Daily Enterprise*, July 21, 1962, JCP.

39. Jacqueline Cochran, "Women in Space: Famed Aviatrix Predicts Women Astronauts within Six Years," *Parade*, April 30, 1961, JCP.

40. Cochran, "Women in Space," 8.

41. Elizabeth Fulbright White to Cochran, May 30, 1961, JCP. Mrs. Carra Harrison had "four sons, aged four to nine years." Carra Elaine Harrison to Cochran, June 3, 1961, JCP.

42. Nancy Lyman to Cochran, May 2, 1961, JCP.

43. Myrtle Thompson Cagle to Cochran, May 1, 1961, JCP; Myrtle Thompson Cagle, interview with author, tape recording, Warner Robins Air Force Base, Robins, Georgia, May 17, 1998. Cochran forwarded letters from "Miss C. Joan Campbell, Mrs. Myrtle Thompson Cagle, Mrs. Marjorie B. Dufton, Mrs. Edward Pirrung, Mrs. Nancy Lynam [*sic*], Fay L. Davis, Miss Carolyn Weinheimer, [and] Ellen Bateman." Floyd Odlum to Dr. W. Randolph Lovelace, May 23, 1961, JCP.

44. Ibid.

45. Lovelace to Fay L. Davis, June 23, 1961, JCP.

46. Floyd Odlum to W. R. Lovelace II, May 31, 1961, JCP.

47. Ibid.

48. Ibid.

49. Cochran to Lovelace, January 14, 1961, JCP.

50. Odlum to Lovelace, February 6, 1961, JCP; Odlum to Lovelace, November 17, 1961, JCP; "Expenses of Medical Checks at Lovelace Clinic (Re: Women 'Astronauts')," JCP.

51. The exact amounts are $18,700 and $55,487.41. Patchy surviving records require comparisons across different years. "Expenses of Medical Checks at Lovelace Clinic (Re: Women 'Astronauts')," JCP; untitled report of fiscal year 1963–64 for the Lovelace Foundation for Medical Research, n.d., Lovelace Inhalation Toxicology Research Institute Library, Kirtland Air Force Base, Albuquerque, New Mexico, 271.

52. Mary Lovelace to Jackie [Cochran] and Floyd [Odlum], January 5, 1949, JCP; Richard G. Elliott, "'On a Comet Always': A Biography of Dr. W. Randolph Lovelace II," *New Mexico Quarterly*, 36 (1966–67): 370; Jackie Lovelace to Cochran, December 25, 1962, JCP.

53. Cochran to Lovelace, August 12, 1949, JCP; telegram, Cochran to Lovelace, Madrid to Albuquerque, October 8, 1955, JCP; Cochran to Lovelace, November 15, 1957, JCP; Lovelace to Cochran, November 29, 1957, JCP; Donald E. Kilgore, oral history interview, edited transcribed tape recording, Albuquerque, New Mexico, April 30, 1997.

54. Lovelace to Cochran, June 9, 1954, JCP; Cochran to Lovelace, January 17, 1964, JCP; Lovelace to Cochran, January 9, 1964, JCP.

55. Lovelace to Odlum, June 8, 1961, JCP, 2.

56. Cochran, *Stars at Noon*, 215–17.

57. Warren G. Woodward, interview by Paul Bolton, June 3, 1968, interview AC 69-82, transcript, Lyndon Baines Johnson Presidential Library, Austin, Texas, 31, 33.

58. Cochran and Bucknum Brinley, *Jackie Cochran*, 316–17.

Chapter 5. Lovelace's Women in Space Program

1. "Damp Prelude to Space: A Potential Lady Orbiter Excels in Lonesome Test," *Life*, October 24, 1960, 81.

2. Wally Funk, oral history interview, edited transcribed tape recording, Trophy Club, Texas, September 24, 1997, 5; "Biographical Information," Papers of Mary Wallace "Wally" Funk II, National Air and Space Museum, Washington, DC (hereafter cited as NASM).

3. Jay T. Shurley to Funk, November 2, 1960, Wally Funk personal collection, Trophy Club, Texas.

4. Jerri Sloan Truhill, oral history interview, edited transcribed tape recording, Richardson, Texas, September 27, 1997, 3; Fran Bera, questionnaire completed for author, January 29, 1998; Lovelace to Frances Bera, September 13, 1960, Fran Bera personal papers, San Diego, California; Lovelace to Frances Bera, September 30, 1960, Fran Bera personal papers, San Diego, California.

5. Lovelace to Sara Lee Gorelick, May 22, 1961, Sarah Gorelick Ratley personal papers, Overland Park, Kansas. Fran Bera received an identical letter on September 13, 1960. Lovelace to Frances Bera, September 13, 1960, Fran Bera personal papers, San Diego, California.

6. "Press Conference, Mercury Astronaut Team," transcript of press conference, April 9, 1959, Historical Collection, National Aeronautics and Space Administration Headquarters History Office, Washington, DC (hereafter cited as NASA); Randy Lovelace, "Duckings, Probings, Checks That Proved Fliers' Fitness," *Life*, April 20, 1959.

7. Lovelace to Sara Lee Gorelick, May 22, 1961.

8. Ibid.

9. Margaret A. Weitekamp, "The Right Stuff, the Wrong Sex: The Science, Culture, and Politics of the Lovelace Woman in Space Program, 1959–1963" (PhD diss., Cornell University, 2001), 215; "'Women in Space'—Girls Tested," n.d., Jacqueline Cochran Papers, Dwight D. Eisenhower Presidential Library, Abilene, Kansas (hereafter cited as JCP); "List of Women Who Completed Medical Tests at Lovelace Foundation, Albuquerque, New Mexico," n.d., JCP.

10. Marilyn C. Link, questionnaire completed for author, May 13, 1998.

11. "'Women in Space'—Girls Tested," n.d., JCP. Irene Leverton, oral history interview, edited transcribed tape recording, Chino Valley, Arizona, November 22, 1997, 6; Martha Ackmann, *The Mercury 13: The Untold Story of Thirteen American Women and the Dream of Space Flight* (New York: Random House, 2003), 77n–78n.

12. Lovelace to Geraldine H. Sloan, March 24, 1961, Jerri Sloan Truhill personal collection, Richardson, Texas. Fran Bera received an identical letter, dated February 28, 1961. Lovelace to Frances S. Bera, February 28, 1961, Fran Bera personal collection, San Diego, California.

13. Ibid.

14. Sarah Gorelick Ratley, oral history interview, edited transcribed tape recording, Overland Park, Kansas, June 7, 1997, 8.

15. "Information from Mrs. Quarles re Women Pilots," n.d., JCP.

16. Virginia Holmes, telephone interview, December 15, 1997.

17. Truhill oral history interview, 3.

18. Ratley oral history interview, 6.

19. Patricia K. Jetton, questionnaire completed for the author, January 14, 1998.

20. Georgiana T. McConnell, questionnaire completed for the author, December 2, 1997; Leverton oral history interview, 6.

21. Bernice Trimble Steadman, oral history interview, edited transcribed tape recording, Traverse City, Michigan, October 8, 1997, 8; Gene Nora Stumbough Jessen, oral history interview, edited transcribed tape recording, Boise, Idaho, May 24, 1997, 6.

22. Jane B. Hart, oral history interview, edited transcribed tape recording, Mackinac City, Michigan, October 7, 1997, 3.

23. For more on the Lovelace Clinic during the late 1950s and early 1960s, see Jake W. Spidle Jr., *The Lovelace Medical Center: Pioneer in American Health Care* (Albuquerque: University of New Mexico Press, 1987), 121–26; Shirley Hassig to Georgiana T. McConnell, March 17, 1961, Georgiana T. McConnell personal papers, Nashville, Tennessee.

24. See also "Table 5: Schedule for the Lovelace Astronaut Screening Program," in Spidle, *Lovelace Medical Center*, 135–36.

25. The Lovelace Foundation originally used this method for the air force's Man in Space program. Working with IBM, the Lovelace physicians recorded information on cardboard punch cards. The astronaut selection process began using pencil-marked cards that the computer could punch automatically while tallying the findings. A. H. Schwichtenberg, Donald D. Flickinger, and W. Randolph Lovelace II, "Medical Machine Record Cards: Their Development and Use in the Astronaut Selection Program," office copy, Lovelace Foundation for Medical Education and Research Papers, Inhalation Toxicology Research Institute Library, Kirtland Air Force Base, Albuquerque, New Mexico, 11; McConnell questionnaire, 3; untitled schedule of testing at the Lovelace Clinic, Jerrie Cobb Papers, Ninety-Nines International Organization of Women Pilots, National Headquarters, Will Rogers Airport, Oklahoma City, Oklahoma, 1 (hereafter cited as JCobbP).

26. W. Randolph Lovelace II, A. H. Schwichtenberg, Ulrich C. Luft, and Robert R. Secrest, "Selection and Maintenance Program for Astronauts for the National Aeronautics and Space Administration," *Aerospace Medicine* 11 (June 1962): 667.

27. Untitled schedule of testing at the Lovelace Clinic, Georgiana T. McConnell personal papers, Nashville, Tennessee, 1.

28. Untitled schedule of testing at the Lovelace Clinic, JCobbP; untitled schedule of testing at the Lovelace Clinic, Georgiana T. McConnell personal papers, Nashville, Tennessee; untitled schedule of testing at the Lovelace Clinic, Fran Bera personal papers, San Diego, California; Gene Nora Stumbough to Mr. and Mrs. Stumbough, July 27, 1961, Gene Nora Stumbough Jessen personal papers, Boise, Idaho, 1.

29. "Instructions for Recording and Using the Basal Temperature Graph," Sarah Ratley personal collection, Overland Park, Kansas; "Basal Temperature Chart," ibid.; Ratley oral history interview, 19–20.

30. Lovelace et al., "Selection and Maintenance Program for Astronauts," 672, 675.

31. Truhill oral history interview, 6; Stumbough to Stumbough, July 27, 1961, 1, 2.

32. Georgiana McConnell questionnaire, 3.

33. Lovelace et al., "Selection and Maintenance Program for Astronauts," 673; Marion Dietrich, "First Woman in Space," *McCall's*, September 1961, 180.

34. Jessen oral history interview, 10.

35. Jetton questionnaire, 4.

36. Stumbough to Stumbough, July 27, 1961; Steadman oral history interview, 9–10; Truhill oral history interview, 5.

37. Steadman oral history interview, 8.

38. Jessen oral history interview, 12; Funk oral history interview, 6; Donald E. Kilgore, oral history interview, edited transcribed tape recording, April 30, 1997, Albuquerque, New Mexico, 15.

39. Lovelace to Jerri Sloan, May 17, 1961, Jerri Sloan Truhill personal papers, Richardson, Texas.

40. Holmes remains claustrophobic ever since this experience. Virginia Holmes, telephone interview, December 17, 1997; Lovelace to McConnell, May 22, 1961, Georgiana T. McConnell personal papers, Nashville, Tennessee.

41. Jessen oral history interview, 13; Bera questionnaire, 3.

42. "After thirty-seven years I have still had no problem." Jetton questionnaire, 4.

43. Jessen oral history interview, 7.

44. Funk oral history interview, 13.

45. Jerrie Cobb to Wally Funk, July 24, 1961, Wally Funk personal collection, Trophy Club, Texas; "Record of Tests Taken," Funk Papers, NASM; Funk oral history interview, 13.

46. Shurley to Funk, November 2, 1960, Wally Funk Personal Collection, Trophy Club, Texas; Shurley to Funk, August 10, 1961.

47. For Funk's own description, see "The Mercury 13 Story," online at www.ninety-nines.org/mercury.html (July 20, 2004). Funk oral history interview, 9; "Record of Tests Taken," Wally Funk Papers, NASM.

48. Funk oral history interview, 17–18.

49. Jerrie Cobb testifying before the Special Subcommittee on the Selection of Astronauts of the Committee on Science and Astronautics, U.S. House of Representatives, 87th Cong., 2nd sess., July 17–18, 1962, 11.

50. Cobb to F.L.A.T. (fellow lady astronaut trainee), May 29, 1961, Jerri Sloan Truhill personal collection, Richardson, Texas.

51. "Release," August 10, 1961, Sarah Gorelick Ratley personal collection, Overland Park, Kansas; Ackmann, *Mercury 13*, 127n.

52. Lovelace to Sloan, July 8, 1961, Jerri Sloan Truhill personal collection, Richardson, Texas.

53. Lovelace to Gorelick, July 12, 1961, Sarah Gorelick Ratley personal collection, Overland Park, Kansas; Lovelace to Sloan, July 12, 1961, Jerri Sloan Truhill personal collection, Richardson, Texas.

54. "Twelve Women Test for Space," *Washington Post*, January 27, 1961, NASM.

55. Marion Dietrich, "First Woman in Space," *McCall's*, September 1961, 80, 81, 184.

56. Cobb to A. B. Smith, August 10, 1961, Sarah Gorelick Ratley personal collection, Overland Park, Kansas.

57. Cobb to Gorelick, July 25, 1961, Sarah Gorelick Ratley personal collection, Overland Park, Kansas.

58. Ratley oral history interview, 7.

59. Cagle to Cochran, July 21, 1961, JCP.

60. Jessen oral history interview, 7.

61. Lovelace to Myrtle Cagle, August 21, 1961, JCP; telegram, Lovelace to Gorelick, September 12, 1961, Sarah Gorlick Ratley personal collection, Overland Park, Kansas. Also, telegram, Lovelace to Funk, September 16, 1961 [*sic*], Wally Funk Papers, NASM.

Chapter 6. Jerrie Cobb, NASA, and the Pensacola Cancellation

1. Doyle to Gorelick, telegram, September 12, 1961, Sarah Gorlick Ratley personal papers, Overland Park, Kansas.

2. W. Henry Lambright, *Powering Apollo: James E. Webb of* NASA (Baltimore: Johns Hopkins University Press, 1995), 82–83, 84.

3. Roger D. Launius, *NASA: A History of the U.S. Civil Space Program* (Malabar, FL: Krieger, 1994), 44.

4. Emphasis added. Memorandum, John F. Kennedy to Vice President Lyndon Johnson, April 20, 1961, John F. Kennedy Presidential Library, Boston (hereafter cited as JFK).

5. Lyndon Baines Johnson to John F. Kennedy, "Evaluation of Space Program," April 28, 1961, Historical Collection, National Aeronautics and Space Administration Headquarters History Office, Washington, DC (hereafter cited as NASA).

6. Lambright, *Powering Apollo*, 96–97; John F. Kennedy, "Excerpts from 'Urgent National Needs,'" Speech to a Joint Session of Congress, May 25, 1961, included in John Logsdon, ed., *Exploring the Unknown: Selected Documents in the History of the U.S.*

Civil Space Program, vol. 1, *Organizing for Exploration*, NASA History Series (Washington, DC: National Aeronautics and Space Administration, 1995), 453.

7. Robert D. Dean, "Masculinity as Ideology: John F. Kennedy and the Domestic Politics of Foreign Policy," *Diplomatic History* 22 (Winter 1998): 30; Robert D. Dean, *Imperial Brotherhood: Gender and the Making of Cold War Foreign Policy* (Amherst: University of Massachusetts Press, 2001).

8. *Washington Star*, October 19, 1975, clipping in "Leonov, Post: ASTP [Apollo Soyuz Test Project]," NASA; Bettyann Holtzmann Kevles, *Almost Heaven: The Story of Women in Space* (New York: Basic Books, 2003), 41; Alan Shepard and Deke Slayton with Jay Barbree and Howard Benedict, *Moon Shot: The Inside Story of America's Race to the Moon* (Atlanta: Turner, 1994), 61.

9. "Biographical Summary," in Jerrie Cobb, "Women's Participation in Space Flight," Aviation/Space Writers Association Annual Meeting, May 1, 1961, Jerrie Cobb Papers, Ninety-Nines International Organization of Women Pilots Headquarters, Will Rogers Airport, Oklahoma City, Oklahoma, 1, 4 (hereafter cited as JCobbP).

10. Cobb, "Women's Participation," 3.

11. Jerrie Cobb to James E. Webb, May 16, 1961, JCobbP.

12. Martha Ackmann, *The Mercury 13: The Untold Story of Thirteen American Women and the Dream of Space Flight* (New York: Random House, 2003), 123–24; Maura Phillips Mackowski, "Human Factors: Aerospace Medicine and the Origins of Manned Space Flight in the United States" (PhD diss., Arizona State University, 2002).

13. "Girl Astronaut?" *New York Herald Tribune*, May 28, 1961, NASA. See also "Jerri Cobb Is Set for Space," *Washington Post*, May 30, 1961, Jacqueline Cochran Papers, Dwight D. Eisenhower Presidential Library, Abilene, Kansas (hereafter cited as JCP).

14. Jerrie Cobb to James E. Webb, May 30, 1961, JCobbP.

15. Emphasis in original. Jerrie Cobb, untitled recommendations to NASA, June 15, 1961, NASA, 3.

16. Ibid., 5.

17. Dr. G. Dale Smith to Mr. George Low, Letter re: Women Astronauts, June 19, 1961, NASA.

18. Cochran to Pirie, August 3, 1961, JCP; Cochran, "Personal and Confidential Memorandum: Subject—Women Astronauts," August 1, 1961, JCP.

19. Jerrie Cobb, oral history interview, edited transcribed tape recording, Denver, Colorado, March 13, 1998, 9; Cobb to Welch, November 16, 1961, JCobbP.

20. Hugh L. Dryden to Vice Admiral R. B. Pirie, October 2, 1961, Dryden Chronological Files, NASA.

21. Donald E. Kilgore, oral history interview, edited transcribed tape recording, Albuquerque, New Mexico, April 30, 1997, 10.

22. Telegram, Jeanne Williams to Jacqueline Cochran, n.d. [September 12, 1961], JCP.

23. W. Randolph Lovelace II to James E. Webb, September 29, 1961, JCP.

24. Ibid., 2.

25. Webb to Cobb, December 15, 1961, JCobbP.

26. Cobb to FLATS, February 12, 1962, Sarah Gorelick Ratley personal collection, Overland Park, Kansas.

27. Ibid.

28. E. K. Hopper, "Lots of Room in Space for Women," *American Girl*, December 1961, n.p., JCP; press release, February 19, 1962, Vice Presidential Papers, Lyndon B. Johnson Presidential, Library, Austin, Texas (hereafter cited as LBJ). Margaret Rossiter documented similar appeals to female scientists. See chapter 3, "'Scientific Womanpower': Ambivalent Encouragement," in *Women Scientists in America: Before Affirmative Action, 1940–1972*, by Margaret W. Rossiter (Baltimore: Johns Hopkins University Press, 1995).

29. Catherine Smith to Johnson, March 15, 1962, Vice Presidential Papers, LBJ; Fritzi Mann to Johnson, March 22, 1962, Vice Presidential Papers, LBJ.

30. Cobb to FLATS, n.d., Jerri Truhill personal collection, Richardson, Texas, 1; Isabelle Shelton, "Capital Footnotes: Senate Wife Could Be First Woman in Space," *Sunday Washington (DC) Star*, March 11, 1962, Vice Presidential Papers, LBJ; outprint of *Congressional Record*, n.d., JCobbP.

31. G. E. R. to Vice President, March 8, 1962, LBJ; "Memo on Requested Meeting with the Vice President," March 7, 1962, LBJ; Jane B. Hart, oral history interview, edited transcribed tape recording, Mackinac City, Michigan, October 7, 1997, 4–5, 20.

Chapter 7. Jerrie Cobb, John Glenn, and the House Subcommittee Hearings

1. Lyndon B. Johnson to James E. Webb, unmailed letter, March 15, 1962, Vice Presidential Papers, Lyndon B. Johnson Presidential Library, Austin, Texas (hereafter cited as LBJ).

2. Memorandum, Liz [Carpenter] to Vice President Johnson, March 14, 1962, Vice Presidential Papers, LBJ.

3. Johnson to Webb, March 15, 1962, LBJ. That evening Johnson had cocktails at Webb's Washington, DC, home. Pre-presidential Daily Diary, March 15, 1962, LBJ.

4. Mrs. George B. Ward Jr. to Johnson, n.d., Vice Presidential Papers, LBJ.

5. Marjorie N. White to Johnson, March 13, 1962; Miss Sue Ann Winkelman to Johnson, March 21, 1962; Johnson to White, March 24, 1962; and Johnson to Winkelman, March 28, 1962, all Vice Presidential Papers, LBJ.

6. Ellipses in original. Cobb to FLATS, n.d., Jerri Sloan Truhill personal collection, Richardson, Texas.

7. Bart Slattery, "Memorandum to Dr. von Braun," May 2, 1962, Wernher von Braun Papers, Manuscripts Division, Library of Congress. Thank you to Bettyann Holtzmann Kevles.

8. Ibid.; von Braun to Cobb, May 19, 1962, Wernher von Braun Papers, Manuscripts Division, Library of Congress.

9. Robert R. Gilruth to Jerrie Cobb, [April 17, 1962], Special Collections: Robert Sherrod Apollo Collection, Historical Collection, National Aeronautics and Space Administration Headquarters History Office, Washington, DC (hereafter cited as NASA).

10. Jane B. Hart, oral history interview, edited transcribed tape recording, Mackinac

City, Michigan, October 7, 1997, 4; Cobb to FLATs, n.d., Jerri Sloan Truhill personal collection, Richardson, Texas.

11. Hart oral history interview, 5.

12. Odlum to Lovelace, May 31, 1961, Jacqueline Cochran Papers, Dwight D. Eisenhower Presidential Library, Abilene, Kansas, 2 (hereafter cited as JCP).

13. Cochran to Lovelace, June 16, 1961, JCP, 2.

14. Cochran to Cagle, July 12, 1961, JCP.

15. Cochran to Cobb, March 23, 1962, JCP, 4.

16. "List of Those to Whom Copies of the Jerrie Cobb Letter Are to Be Sent," n.d., JCP; James E. Webb to Jacqueline Cochran, July 12, 1962, JCP. Cochran to Hugh L. Dryden, June 26, 1962, JCP; Cochran to Robert Gilruth, June 18, 1962, JCP; Cochran to General Curtis Le May, June 15, 1962, JCP.

17. Lovelace to Cochran, June 29, 1962, JCP, 2.

18. D. Brainard Holmes, "Memorandum for Associate Administrator, Subject: Views on Women Astronauts," June 7, 1962, NASA.

19. James E. Webb, Speech before the General Federation of Women's Clubs, Washington, DC, June 27, 1962, NASA, 12.

20. Hart oral history interview, 6.

21. Jo Freeman, "How 'Sex' Got into Title VII: Persistent Opportunism as a Maker of Public Policy," *Law and Inequality: A Journal of Theory and Practice* 9 (March 1991): 163–84.

22. Congress, House, Committee on Science and Astronautics, *Qualifications for Astronauts: Hearings before the Special Subcommittee on the Selection of Astronauts of the Committee on Science and Astronautics*, 87th Cong., 2nd sess., July 17–18, 1962, 7.

23. Ibid., 5.

24. Ibid., 14, 18.

25. Ibid., 19.

26. Ibid., 14, 24.

27. Ibid., 28, 27, 38. Cochran took liberties with this figure, attributing all washouts to voluntary drops for marriage. The WASP 40 percent washout figure paralleled the statistics for men's aviation training.

28. Joseph Hearst, "Give Us Space Role, Women Pilots Urge," *Chicago Tribune*, July 18, 1962, NASA; "Of Sex and Spaceniks: Cochran Briefs Congress," *New York Daily News*, July 18, 1962, NASA; Women, War, and Headline News conference, Women in Military Service for America Memorial, Arlington National Cemetery, Arlington, Virginia, March 7, 2000.

29. Congress, House, *Qualifications for Astronauts: Subcommittee Hearings*, 54; Nina E. Lerman, "'Preparing for the Duties and Practical Business of Life': Technological Knowledge and Social Structure in Mid-19th-Century Philadelphia," in special issue, "Gender Analysis and the History of Technology," *Technology and Culture* 38 (January 1997): 36.

30. Wally Funk, oral history interview, edited transcribed tape recording, Trophy Club, Texas, September 24, 1997, 3.

31. Congress, House, *Qualifications for Astronauts: Subcommittee Hearings*, 53.

32. Ibid., 55, 64.

33. Ibid., 67.

34. "It [the cancellation] was just such a shock. We were all packed, ready to go to Pensacola. So thrilled. We were finally going to get our hands on a jet." Jerri Sloan Truhill, oral history interview, edited transcribed tape recording, Richardson, Texas, September 27, 1997, 7.

35. Tom Wolfe, *The Right Stuff* (New York: Farrar, Straus, Giroux, 1979); Walter Cunningham with Mickey Herskowitz, *The All-American Boys* (New York: Macmillan, 1977), 245.

36. Congress, House, *Qualifications for Astronauts: Subcommittee Hearings*, 74–75.

37. Ibid., 75, 59.

38. George P. Miller to Cochran, July 30, 1962, JCP; James E. Webb, July 18, 1962, Daily Engagement Books, James E. Webb Papers, Harry S. Truman Presidential Library, Independence, Missouri.

39. Congress, House, *Qualifications for Astronauts: Subcommittee Hearings*, 9.

40. Sarah Gorelick Ratley, oral history interview, edited transcribed tape recording, Overland Park, Kansas, June 7, 1997; Hart interview, October 7, 1997; Bernice "B" Steadman, oral history interview, edited transcribed tape recording, Traverse City, Michigan, October 8, 1997; Gene Nora Stumbough Jessen, oral history interview, edited transcribed tape recording, Boise, Idaho, May 24, 1997.

41. Cobb to the President, telegram, July 20, 1962, NASA; Webb to Cobb, August 3, 1962, NASA; Cobb to Webb, August 7, 1962, Jerrie Cobb Papers, Ninety-Nines International Organization of Women Pilots Headquarters, Will Rogers Airport, Oklahoma City, Oklahoma (hereafter cited as JCobbP).

42. Jerrie Cobb, "Project WISE," Space Symposium, Air Force Association's Sixteenth National Convention and Aerospace Panorama, Las Vegas, Nevada, September 21, 1962, JCobbP, 4.

43. Jerrie Cobb as quoted in *Washington Post*, November 16, 1962, typewritten excerpts, Jerrie Cobb files, NASA; Cobb, "Women in Space," n.d., JCobbP; Zonta Club of Cleveland, "Women in the Space Age," dinner program, November 28, 1962, JCobbP.

Chapter 8. Several Epilogues to Lovelace's Women in Space Program

1. Memorandum, Robert P. Young, "Memorandum for Record; Subject: Meeting between the Administrator and Miss Jerrie Cobb," December 27, 1962, Young, R. P., Memos (1961–62), NASA Historical Collection, National Aeronautics and Space Administration Headquarters History Office, Washington, DC (hereafter cited as NASA).

2. Ibid., 2.

3. Asif A. Siddiqi, *Challenge to Apollo: The Soviet Union and the Space Race, 1945–1974* (Washington, DC: National Aeronautics and Space Administration, NASA History Division, Office of Policy and Plans, 2000), 370.

4. Margaret Chase Smith, "Tape on Women in Space," Business and Professional Women's Luncheon, January 19, 1963; and Margaret Chase Smith, interview by Mar-

tin Agronsky, NBC-TV-Cooper, May 6, 1963 (for May 13, 1963), both in Margaret Chase Smith Library, Northwood University, Skowhegan, Maine.

5. Robert B. Voas, "Speech to Downtown YMCA," February 1, 1963, Robert Voas Papers, NASA.

6. Ibid.

7. Cobb to John F. Kennedy, March 13, 1963, John F. Kennedy Presidential Library, Boston, Massachusetts (hereafter cited as JFK).

8. Webb to Cobb, April 13, 1963, Jerrie Cobb Files, NASA.

9. Central Intelligence Agency Information Report, "Reported Plans for a Soviet Space Spectacular," March 26, 1963, Vice Presidential Security File, Lyndon Baines Johnson Presidential Library, Austin, Texas (hereafter cited as LBJ).

10. Valentina Ponomareva and Debra D. Facktor, "The Flight That Never Happened: The Story of the First Women Cosmonaut Team," Forty-seventh International Astronautical Congress, October 7–11, 1996, Beijing, China, NASA, 3, 6–7; Siddiqi, *Challenge to Apollo,* 365. The other four female cosmonauts remained in the Soviet space program until the Central Committee denied their 1969 letter requesting to serve in space.

11. In contrast, the last Project Mercury solo launch completed twenty-two orbits in thirty-four hours and twenty minutes on May 16, 1963. *NASA Pocket Statistics* (Washington, DC: National Aeronautics and Space Administration, Headquarters Facilities and Logistics Management, 1997), B-94.

12. William Shelton, *Soviet Space Exploration: The First Decade* (New York: Washington Square Press, 1968), 159; A. Lothian, *Valentina: First Woman in Space, Conversations with A. Lothian* (Edinburgh, UK: Pentland Press, 1993).

13. James J. Harford, *Korolev: How One Man Masterminded the Soviet Drive to Beat America to the Moon* (New York: John Wiley, 1997), 179; William Burroughs, *This New Ocean: The Story of the First Space Age* (New York: Random House, 1998).

14. Siddiqi, *Challenge to Apollo,* 371–72. On August 20, 1982, cosmonaut Svetlana Savitskaya took a Soyuz spacecraft to the Soviet *Salyut* space station for an eight-day mission with two male cosmonauts. In 1984 she became the first woman to walk in space. Her launch beat American Sally Ride's June 1983 spaceflight by almost a year and Kathryn Sullivan's U.S. spacewalk by three months.

15. Memorandum, Bart Slattery to Dr. von Braun, June 21, 1963, National Archives and Records Administration, Record Group 255, Atlanta, Georgia. Thank you to Michael Neufeld for this reference.

16. "It Was a Wonderful Space Feat—but Exactly What Did It Prove?" *New York Herald Tribune,* June 20, 1963, NASA.

17. Joy Miller, "Red Space Girl Irks U.S. Woman Flyer," *Marshall (TX) News Messenger,* July 7, 1963, Jacqueline Cochran Papers, Eisenhower Presidential Library, Abilene, Kansas (hereafter cited as JCP); Joy Miller, "Space-Bitten U.S. Gal Chagrined," *Hackensack (NJ) Record,* June 17, 1963, JCP; Drew Pearson, "Woman Pilots Angry at Webb," *Washington Post,* June 20, 1963, NASA; Claire Wallace, "Spacewoman up in the Air over Red Tape," *New York World-Telegram and Sun,* June 21, 1963, JCP.

18. Jerrie Cobb and Jane Rieker, *Woman into Space: The Jerrie Cobb Story* (Englewood Cliffs, NJ: Prentice-Hall, 1963); Louise Sweeney, "Why Valentina and Not Our Gal?" *Pittsfield (MA) Berkshire Eagle*, June 21, 1963, NASA.

19. Clare Boothe Luce, "A Monthly Commentary," *McCall's*, May 1963, Jerrie Cobb Papers, Ninety-Nines International Organization of Women Pilots Headquarters, Will Rogers Airport, Oklahoma City, Oklahoma (hereafter cited as JCobbP).

20. Clare Boothe Luce, "But Some People Simply Never Get the Message," *Life*, June 28, 1963, 31. The magazine's cover featured a banner reading, "Clare Boothe Luce, Soviet Space Girl Makes U.S. Men Sound Stupid."

21. Luce, "But Some People Simply Never Get the Message."

22. "The U.S. Team Is Still Warming Up the Bench," *Life*, June 28, 1963, 32–33.

23. Edith Hills Coogler, "Ladies, Man Your Ship! Macon Astronette Stands By," *Atlanta Journal*, June 26, 1963, JCP; Edith Hills Coogler, "Comments on Space Trip: Myrtle Thompson Cagle: 'It Should Have Been Me,'" *Raleigh (NC) News and Observer*, July 9, 1963, JCP; Coogler, "Possible Woman Astronaut: Tar Heel Sets Sights on Moon," *Charlotte (NC) Observer*, July 14, 1963, JCP.

24. Johnnie R. Betson Jr., M.D. and Robert R. Secrest, M.D., "Prospective Women Astronauts Selection Program: Rationale and Comments," *American Journal of Obstetrics and Gynecology* 88 (February 1, 1964): 421–23.

25. Betson and Secrest, "Prospective Women Astronauts Selection Program," 421, 422.

26. Betson and Secrest, "Prospective Women Astronauts Selection Program," 423.

27. "A Space Official Missing on Flight; Dr. Lovelace Is Unreported with His Wife and Pilot," *New York Times*, December 14, 1965; Richard G. Elliott, "'On A Comet Always': A Biography of Dr. W. Randolph Lovelace II," *New Mexico Quarterly* 36 (1966–67): 351–52, 381–83.

28. Donald E. Kilgore, oral history interview, edited transcribed tape recording, Albuquerque, New Mexico, April 30, 1997, 21; Jake W. Spidle Jr., *The Lovelace Medical Center: Pioneer in American Health Care* (Albuquerque: University of New Mexico Press, 1987), 142, 143. See especially chapter 6, "A New Era."

29. Spidle, *Lovelace Medical Center*, 140 n. 34; "Lovelace's Dedication to Man Eulogized by Space Official," *Albuquerque Tribune*, March 1, 1966, JCP; "Space Medicine Leader, His Wife and Pilot Found Dead in Plane Wreck," *New York Times*, December 16, 1965.

30. Memorandum, Jacqueline Cochran to Hon. James Webb, Dr. Hugh Dryden, re: Women in Space, July 26, 1962, JCP, 2–3.

31. Response to Inquiry from Congress, July 9, 1963, Jacqueline Cochran File, NASA; press release, Office of the Administrator, NASA, n.d., JCP; Appointment Affidavits, June 11, 1963, JCP; Dryden to Cochran, January 22, 1963, JCP.

32. Cochran to Webb, July 2, 1963, JCP, 2; Cochran to Webb, July 3, 1963, JCP; Cochran to Lovelace, July 2, 1964, JCP.

33. Emphasis in original. [Jacqueline Cochran], untitled plan for testing "if women could operate in space," n.d. [1967], JCP.

34. Cochran to Webb, June 17, 1965, JCP; Robert R. Gilruth to Cochran, October 26, 1966, JCP; Julian Scheer to Cochran, telegram, n.d. [1965], JCP; Cochran to Webb, August 17, 1965, JCP; invitation list, November 29, 1968, White House Central Files, LBJ.

35. C. J. George to Cochran, July 15, 1969, JCP; Cochran to Dr. T. O. Paine, August 8, 1969, NASA.

36. John W. Macy Jr., "Memorandum for Mr. Jack Valenti, Subject: Jacqueline Cochran," April 20, 1965, White House Central Files, Name Files, LBJ; MBC, "Memo for the Record," January 31, 1967, Office Files of John Macy, LBJ; Kilgore oral history interview, 16, 17.

37. Kilgore oral history interview, 16.

38. "Aviatrix Applies for Space Training but Is Too Late," *Newport News (VA) Daily Press*, July 13, 1963, NASA.

39. Frank Macomber, "No Room in Space for Gal Astronaut," *Joliet (IL) Herald News*, June 12, 1963, JCP.

40. Cobb to the president, February 10, 1964, NASA; Cobb to Clinton R. Anderson, February 11, 1964, NASA; Lyndon Johnson to Cobb, March 17, 1964, NASA.

41. Jerrie Cobb, *Jerrie Cobb: Solo Pilot* (Sun City Center, FL: Jerrie Cobb Foundation, 1997), 163–78.

42. Gene Nora Jessen, "Spurned by NASA Female Pilot Turns to Flying Amazonia," clipping attached to Jerrie Cobb Foundation newsletter, November 14, 1973, JCP.

43. "Oklahoma Woman Honored for South American Flights," *New York Times*, October 16, 1972, NASA; newspaper clipping of photograph and caption, *Washington Post*, September 21, 1973, NASA.

44. Irene Leverton to author, n.d. [1997]; Sarah Gorelick Ratley, oral history interview, edited transcribed tape recording, Overland Park, Kansas, June 7, 1997, 12; Gene Nora Stumbough Jessen, May 24, 1997, oral history interview, edited transcribed tape recording, Boise, Idaho, 9, 15.

45. Gene Nora Jessen, *The Powder Puff Derby of 1929: The First All-Women's Transcontinental Air Race* (Naperville, IL: Sourcebooks Trade, 2002).

46. Rhea Woltman, questionnaire completed for author, May 18, 1999, 3.

47. Jane B. Hart, oral history interview, edited transcribed tape recording, October 7, 1997, Mackinac City, Michigan, 6.

48. Jeanette Jenkins, "Personality Profile," June 1985, Jean Hixson clippings collection, originals held by Pauline Vincent, Glenwood Springs, Colorado.

49. Joan McCullough, "The 13 Who Were Left Behind," *Ms.*, September 1973, 41–45.

50. Interview with Lt. Col. Eileen Collins, September 1, 1999, New York State Fair, Syracuse, New York. Thank you to Emily Longnecker, John Herrick, and WENY Channel 36 Elmira, New York, for this opportunity.

51. All twenty-five names are in Margaret A. Weitekamp, "The Right Stuff, the Wrong Sex: The Science, Culture, and Politics of the Lovelace Woman in Space Program, 1959–1963" (PhD diss., Cornell University, 2001), 215.

52. Collins interview, Syracuse, New York, September 1, 1999.

53. Wally Funk, oral history interview, edited transcribed tape recording, September 24, 1997, Trophy Club, Texas, 5. Interorbital Systems can be found online at www.interorbital.com; Trans Lunar Research can be found online at www.translunar.org.

54. "The Wrong Stuff?" *Dateline NBC*, February 10, 1995; Susan Carpenter, "Rocket Grrrls!" *George*, September 1997, 136–39, 145. The origin of the name "Mercury Thirteen" was sharply contested in 1999. James M. Cross, "Legend of the 'Mercury 13'—Confessions of a TV Journalist," *International Women Pilots/99s News*, January–February 1999, 7–9; "Letters to the Editor," *International Women Pilots/99s News*, March–April 1999.

55. Ben Hellwarth, "If John Glenn, Why Not the Women? UCSB Professor Suggests Females Who Trained with the Mercury Astronauts Be Given an Opportunity," *Santa Barbara News-Press*, January 29, 1998.

56. Jerrie Cobb, oral history interview, edited transcribed tape recording, conducted at the Women in Aviation, International conference in Denver, Colorado, March 13, 1998, 16–17.

57. See Cobb, *Solo Pilot*, chapter 11, "Flight to Space? Year 1959–1963, Age 28–32," 143–60.

58. Paul Hoverman, "Letters to NASA: 'Fly Me': Space-Seekers Want to Float in Glenn's Shoes," *USA Today*, July 15, 1998; conversation with Jennifer McCarter, Code P [Public Relations], NASA Headquarters, Washington, DC, July 2, 1998; Marcia Dunn, "NASA Pioneer Asks for Her Shot at Space: 1st Female Astronaut Hopeful Eyes Shuttle," *Washington Post*, July 13, 1998.

59. Conversation with Jennifer McCarter, Code P, NASA Headquarters, Washington, DC, July 2, 1998.

60. "*Flyer* Advertisement," *Facts and Figures: The Official Newsletter of the Six Figures Theatre Company* 2, no. 1, 1.

Conclusion

1. Betty Friedan, *The Feminine Mystique* (New York: W. W. Norton, 1997).

2. Susan J. Douglas, *Where the Girls Are: Growing Up Female with the Mass Media* (New York: Times Books, 1994), 125; Sara Evans, *Personal Politics: The Roots of Women's Liberation in the Civil Rights Movement and the New Left* (New York: Random House, 1980); Ruth Rosen, *The World Split Open: How the Modern Women's Movement Changed America* (New York: Penguin Books, 2000).

3. Margaret W. Rossiter, *Women Scientists in America: Before Affirmative Action, 1940–1972* (Baltimore: Johns Hopkins University Press, 1995), 68.

4. Douglas, *Where the Girls Are*, 125.

5. For a recent edition, see Boston Women's Health Book Collective, *"Our Bodies, Ourselves" for the New Century: A Book by and for Women* (New York: Simon and Schuster, 1998).

6. The Ladies Professional Golf Association began in 1950 and the Women's Tennis Association in 1971. Currently, the WNBA (Women's National Basketball Association) and the Women's United Soccer Association play regularly, as do less-publicized

leagues such as the Women's Professional Football League, in its fifth season in 2003. Lissa Smith et al., eds., *Nike Is a Goddess: The History of Women in Sports* (New York: Atlantic Monthly Press, 1999); HBO Sports, *Dare to Compete: The History of Women in Sports*, documentary film, 1999.

7. Harold Sandler, "Physiological Responses of Women to Simulated Weightlessness: Review of the Significant Findings of the First Female Bedrest Study," Washington, DC, Scientific and Technical Information Office, National Aeronautics and Space Administration, 1978.

8. For an account of the 1970s selection, see Joseph D. Atkinson and Jay M. Shafritz, *The Real Stuff: A History of NASA's Astronaut Recruitment Program* (New York: Praeger, 1985), especially chapter 6, "Selection of Pilots and Mission Specialists (Group VIII)," 133–79.

9. Georgiana T. McConnell to author, July 28, 1999, 5.

10. "Retired Woman Pilot Remembers Wanting to Pilot Space," segment 4, National Public Radio's *Weekend Edition/Saturday* transcript, original airdate February 11, 1995, 7.

ESSAY ON SOURCES

Researching this book required years of work to integrate hard-to-find documentary sources with oral history interviews conducted with the participants. Eleven original fully transcribed, participant-edited oral history interviews provided crucial insights into what it was like both to take Lovelace's examinations and to fly as a female pilot in the 1950s. Seven other women pilots, including NASA astronaut Eileen Collins, either completed questionnaires for this work or submitted graciously to less intensive interviews. In addition to the oral histories conducted especially for this book, oral histories from two other collections provided valuable information. First, the New Mexico Medical History Program at the Medical Center Library of the University of New Mexico in Albuquerque contains three helpful oral history interviews with physicians from the Lovelace Foundation. Second, the Columbia University Oral History Collection holds oral histories conducted with Dr. William Randolph Lovelace II and pioneering aviator Ruth Nichols.

No single collection of documents about Lovelace's Woman in Space Program exists. Rather, evidence about this forestalled program remains in collections across the country. Collectively, the women who participated still hold some of the best sources: the original copies of the correspondence they received. In addition, many kept clippings of contemporary newspaper and magazine articles. Their generosity in allowing me to see these privately held materials made this book possible.

Formal archives also held pieces of the puzzle. The Aeronautics Archives at the Smithsonian Institution's National Air and Space Museum in Washington, DC, maintain individual files on many of the participants as well as an extensive collection of materials about other relevant women pilots. Unfortunately, the extensive Lovelace archives that Jake W. Spidle Jr. assembled when writing his excellent institutional history of the Lovelace Medical Center (listed below) survive only in a severely depleted form in the library at the Inhalation Toxicology Research Institute on the Kirtland Air Force Base near Albuquerque, New Mexico. The collection includes an extensive bibliography of Lovelace Foundation publications, although not the original articles. The NASA Historical Reference Collection at the NASA Headquarters History Office in

Washington, DC, holds files on Cochran, Cobb, Lovelace, Tereshkova, and Webb as well as a growing file on Women in Space and ample material about the United States space effort. Because Lovelace's efforts never became a NASA program, however, the space agency does not have any documents from the program itself.

The Pre-presidential Papers and Presidential Papers at the John F. Kennedy Presidential Library in Boston, Massachusetts, offer insights into the development of space policy during the Kennedy administration, and the Vice-Presidential Papers at the Lyndon Baines Johnson Presidential Library in Austin, Texas, hold documents regarding Johnson's work as head of the Space Council. NASA administrator James Webb's papers at the Harry S. Truman Presidential Library in Independence, Missouri, contain some correspondence with Cochran but little of direct relevance to this history. The archives of the National Federation of Business and Professional Women's Clubs, Inc., in Washington, DC, have the texts of some relevant talks given to its member organizations. The Margaret Chase Smith Library at Northwood College in Skowhagen, Maine, and the Women Airforce Service Pilots Papers in the Woman's Collection at Texas Woman's University in Denton, Texas, sent copies of the texts from several speaking engagements and personal letters, respectively.

Jerrie Cobb deposited some of her personal papers with the Ninety-Nines, the International Organization of Women Pilots, which maintains a small archive at the organization's national headquarters at Will Rogers Airport in Oklahoma City, Oklahoma. She also claims to hold a trunkful of relevant materials that no researcher I know has managed to see. Even though Jane Rieker wrote most of it without Cobb's personal input, Jerrie Cobb and Jane Rieker, *Woman into Space: The Jerrie Cobb Story* (Englewood Cliffs, NJ: Prentice-Hall, 1963), offers a lively and useful account of Cobb's efforts for space. Cobb's more recent autobiography, *Jerrie Cobb: Solo Pilot* (Sun City Center, FL: Jerrie Cobb Foundation,1997), essentially omits the Lovelace episode, offering only a compilation of newspaper clippings as a chapter. The rest of the volume describes Cobb's early aviation career and later missionary efforts.

Sources on Jacqueline Cochran remain simultaneously ample and surprisingly scarce. Her voluminous papers reside at the Eisenhower Presidential Library in Abilene, Kansas. After Radcliffe College representatives approached Cochran in 1951 about depositing her papers in their archives, she kept everything, including incoming and outgoing correspondence, Christmas cards, golf scores, and recipes. Cochran even filed away letters that she herself had labeled "Please destroy." She subscribed to clippings services to collect articles about herself from newspapers and magazines. Thanks to tremendous efforts by the staff at the Eisenhower Library, her papers provide a rich and well-organized source of information about her life, her accomplishments, and her relationships.

At the same time, the two existing "autobiographies" of Cochran present difficulties for researchers. The first, Cochran's own book, *The Stars at Noon* (Boston: Little, Brown, 1954), reads as if you are sitting down with the world-famous flier while she tells you stories as they occur to her. As a result, it remains difficult to establish a definitive chronology. The second, an "autobiography" written by Maryann Bucknum Brinley in collaboration with Cochran just before she died, *Jackie Cochran: An Autobiography* (New York: Bantam Books, 1987), unfolds chronologically but adds other problems from a scholarly perspective. The book integrates Cochran's stories with recollections from her friends and colleagues, presented in sections titled "Other Voices." The remembered accounts do not appear to have been verified using documentary evidence. In addition, the material added is impossible to examine more closely because the recollections remain unaccredited. A scholarly reconsideration of this important twentieth-century woman still awaits an author.

The comparison of Cochran and Earhart in chapter 1 was inspired by Susan Ware's excellent analysis of Earhart, *Still Missing: Amelia Earhart and the Search for Modern Feminism* (W. W. Norton, 1993), and a brief description of an early "aviatrix" in Kathy Peiss, *Hope in a Jar: The Making of America's Beauty Culture* (New York: Metropolitan Books, 1998). Peiss credited pilot Ruth Elder with an action "both forceful and feminine" when, after being rescued from a failed attempt at an Atlantic crossing, she stepped out of her airplane and promptly powdered her nose. I would argue that Elder's nose powdering was a defensive assertion of her femininity in the face of criticism commonly leveled at women pilots. Peiss's arguments prompted me to reconsider Cochran's frequent public application of makeup, however. The comparison emerged with Earhart, persuasively described by Ware as a postsuffrage feminist who eluded contemporary gender stereotypes.

For more on the Collier Trophy see Pamela E. Mack, ed., *From Engineering Science to Big Science: The NACA and NASA Collier Trophy Research Project Winners*, NASA History Series (Washington, DC: National Aeronautics and Space Administration, 1998), and William Robie, *For the Greatest Achievement: A History of the Aero Club of America and the National Aeronautic Association* (Washington, DC: Smithsonian Institution Press, 1993).

For information on the life of William Randolph "Randy" Lovelace II, the best sources remain Richard G. Elliott, "'On a Comet Always': A Biography of Dr. W. Randolph Lovelace II," *New Mexico Quarterly* 36 (1966–67): 351–88, an article written in tribute to the fallen doctor in the year after his death, and Jake W. Spidle Jr., *The Lovelace Medical Center: Pioneer in American Health Care* (Albuquerque: University of New Mexico Press, 1987). Spidle's book is a well-written, engaging history of the two Drs. Lovelace and the Lovelace Foundation and Clinic. Few institutional histories make good summer reading; this one did. For more on physicians and medicine in New Mexico see Jake W. Spidle Jr., *Doctors of Medicine in New Mexico* (Albuquerque: University of New Mexico Press, 1986).

Two dissertations have advanced the historical understanding of aviation and aerospace medicine. Adrianne Noe's "Medical Principle and Aeronautical Practice: American Aviation Medicine to World War II" (PhD diss., University of Delaware, 1989) provides a good overview of how aviation medicine developed during the interwar years as a technology-dependent medical discipline. Maura Phillips Mackowski's "Human Factors: Aerospace Medicine and the Origins of Manned Space Flight in the United

States" (PhD diss., Arizona State University, 2002) takes a wider view, including German scientific explorations and bringing the story up to the 1960s.

One older reference also bears inclusion. In the 1960s, Shirley Thomas compiled a multivolume series based on her interviews with space researchers titled *Men of Space: Profiles of the Leaders in Space Research, Development, and Exploration* (Philadelphia: Chilton Books, 1962). Even forty years later, the lives of these extraordinary men make fascinating reading.

For information on the history of women in aviation see Dean Jaros, *Heroes without Legacy: American Airwomen, 1912–1944* (Niwot: University Press of Colorado, 1993), and Henry Holden and Lori Griffith, *Ladybirds II: The Continuing Story of American Women in Aviation* (Mount Freedom, NJ: Blackhawk, 1993). The Smithsonian Institution Press's series titled *United States Women in Aviation* offers two very good volumes covering 1919–29 (by Kathleen Brooks-Pazmany), and 1930–39 (by Claudia M. Oakes) as well as an extraordinary history by Deborah G. Douglas analyzing the period from 1940 to 1985. One of the women who underwent Lovelace's 1961 astronaut tests has also published a book about a famous 1929 women's aviation race: Gene Nora Jessen, *The Powder Puff Derby of 1929: The First All-Women's Transcontinental Air Race* (Naperville, IN: Sourcebooks Trade, 2002). Numerous books by participants and historians have explored the Women Airforce Service Pilots. One recent example, *Clipped Wings: The Rise and Fall of the Women Airforce Service Pilots (WASPs) of World War II* (New York: New York University Press, 1998), by Molly Merryman, usefully critiques previous historians' fixation with portraying Cochran and Nancy Harkness Love as opposed to each other, characterizing such constructions as ultimately "counterproductive."

Several works helped to flesh out the history of women and gender in a Cold War context. Elaine Tyler May's fundamental work, *Homeward Bound: American Families in the Cold War Era* (New York: Basic Books, 1988), argues that rigidly defined gender roles shaped the context in which people returned to their lives in the postwar period. The articles included in Joanne Meyerowitz, ed., *Not June Cleaver: Women and Gender in Postwar America, 1945–1960*, Critical Perspectives on the Past (Philadelphia: Temple University Press, 1994), broaden the historical understanding of postwar society and politics. As an analysis of postwar popular culture and women, Susan G. Douglas's *Where the Girls Are: Growing Up Female with the Mass Media* (New York: Times Books, 1994) outlined the cultural context in which the issue of women astronauts emerged in the early 1960s.

Two other women's histories provided important frameworks. Margaret Rossiter's comprehensive analysis of female scientists, *Women Scientists in America: Before Affirmative Action, 1940–1972* (Baltimore: Johns Hopkins University Press, 1995) offers an insightful analysis of women's struggles to be accepted in scientific fields during the postwar and Cold War periods. Rossiter's history of postwar women who responded to messages recruiting female scientists—only to encounter institutional barriers—offers an important model for the smaller, parallel history of women presenting themselves as potential astronaut candidates. In considering the ways that Lovelace's Woman in Space

Program and Cobb's battle to resume it serve as harbingers of coming social changes, Ruth Rosen's *The World Split Open: How the Modern Women's Movement Changed America* (New York: Penguin Books, 2000) offered an engagingly detailed history of the political context of the early 1960s and the social movement's origins.

I relied on several excellent space histories. Walter McDougall's . . . *the Heavens and the Earth: A Political History of the Space Age* (New York: Basic Books, 1985) integrates the development of space exploration into its national and international political context, while Roger D. Launius's *NASA: A History of the U.S. Civil Space Program* (Malabar, FL: Krieger, 1994) offers an excellent account of the space agency's origins and history. Joseph D. Atkinson Jr. and Jay M. Shafritz coauthored *The Real Stuff: A History of NASA's Astronaut Recruitment Program* (New York: Praeger, 1985), a history of each astronaut class recruited by NASA, which also includes a chapter on the first attempts by women and minorities to become astronauts. Regarding the Soviet space program, Asif A. Siddiqi's impressive *Challenge to Apollo: The Soviet Union and the Space Race, 1945–1974*, (Washington, DC: National Aeronautics and Space Administration, NASA History Division, Office of Policy and Plans, 2000) is a comprehensive and readable history of the Soviet space effort. The NASA History Office has produced not only impressive histories but also important documentary collections. The multivolume series of NASA documents collectively titled *Exploring the Unknown: Selected Documents in the History of the U.S. Civil Space Program*, published in the NASA History Series and edited by John Logsdon et al., collects invaluable primary sources on the history of the U.S. space program.

The question of women, gender, and space has begun to receive broader attention in recent years. Other volumes addressing various aspects of the history of Lovelace's Woman in Space Program include one account by a participant, *Tethered Mercury* (Traverse City, MI: Aviation Press, 2001), written by Bernice Trimble Steadman and coauthored by Jody M. Clark, as well as another coauthored volume that combines the Lovelace program's history with a history of the Women Airforce Service Pilots: *Amelia Earhart's Daughters: The Wild and Glorious Story of American Women Aviators from World War II to the Dawn of the Space Age*, by Leslie Haynsworth and David M. Toomey (New York: William Morrow, 1998). Stephanie Nolan's *Promised the Moon: The Untold Story of the First Women in the Space Race* (New York: Four Walls Eight Windows, 2002) and Martha Ackmann's *The Mercury 13: The Untold Story of Thirteen American Women and the Dream of Space Flight* (New York: Random House, 2003) also recount the history. Ackmann's chapter 7, "Project Venus," showcases original research on the psychological testing conducted in Oklahoma City.

Recently, two works have begun to remedy the lack of attention to female astronauts. Laura S. Woodmansee published an interactive history of female astronauts titled simply *Women Astronauts* (Burlington, ON: Apogee Books, 2002). Packaged with a CD-ROM, the comprehensive volume, which features information on all women who have flown in space (including the first Soviet, American, British, Canadian, Japanese, and French female astronauts), serves as a needed reference work that would also engage younger readers. Bettyann Holtzmann Kevles has written the first integrated history of female space travelers. *Almost Heaven: The Story of Women in Space* (New York: Basic Books, 2003) examines the historical contexts in which the Soviet Union and the United States integrated women into their respective space programs. Both are welcome contributions.

Finally, several authors have critiqued the gendered politics and culture within and around the United States' human spaceflight program. The first half of Constance Penley's *NASA/TREK: Popular Science and Sex in America* (New York: Verso, 1997) offers an insightful analysis of gender and NASA, focusing particularly well on the 1986 *Challenger* accident. Similarly, in a cultural critique of America's late 1990s obsession with aliens, Jodi Dean's *Aliens in America: Conspiracy Cultures from Outerspace to Cyberspace* (Ithaca: Cornell University Press, 1998) analyzes how the gendering of space itself set the backdrop for the cultural ideas that got projected onto it. Finally, Robert D. Dean's *Imperial Brotherhood: Gender and the Making of Cold War Foreign Policy* (Amherst: University of Massachusetts Press, 2001) demonstrates how gender pervaded foreign policy decision making at the beginning of the space age.

INDEX